ISOELECTRIC FOCUSING

CONTRIBUTORS

GERHARD BAUMANN

JOHAN BOURS

NICHOLAS CATSIMPOOLAS

ANDREAS CHRAMBACH

JOHN S. FAWCETT

ALEXANDER KOLIN

BERTOLD J. RADOLA

HARRY RILBE

OLOF VESTERBERG

COLIN W. WRIGLEY

ISOELECTRIC FOCUSING

Edited by

NICHOLAS CATSIMPOOLAS

Massachusetts Institute of Technology
Cambridge, Massachusetts

ACADEMIC PRESS   New York   San Francisco   London   1976

A Subsidiary of Harcourt Brace Jovanovich, Publishers

ACADEMIC PRESS, INC.
111 Fifth Avenue, New York, New York 10003

United Kingdom Edition published by
ACADEMIC PRESS, INC. (LONDON) LTD.
24/28 Oval Road, London NW1

Library of Congress Cataloging in Publication Data

Main entry under title:

Isoelectric focusing.

    Includes bibliographies.
      1.   Isoelectric focusing.   I.   Catsimpoolas,
Nicholas.   [DNLM: 1.   Isoelectric focusing.
QD79.E44 I851]
QP519.9.I8I78 1976   547'.7'028   75-44759
ISBN 0–12–163950–9

CONTENTS

4   ISOELECTRIC FOCUSING ON POLYACRYLAMIDE GEL

ANDREAS CHRAMBACH AND GERHARD BAUMANN

5   ISOELECTRIC FOCUSING/ELECTROPHORESIS IN GELS

COLIN W. WRIGLEY

6   ISOELECTRIC FOCUSING IN GRANULATED GELS

BERTOLD J. RADOLA

7   CONTINUOUS-FLOW ISOELECTRIC FOCUSING

JOHN S. FAWCETT

8   ISOELECTRIC FOCUSING IN FREE SOLUTION

JOHAN BOURS

9   **TRANSIENT STATE ISOELECTRIC FOCUSING**

NICHOLAS CATSIMPOOLAS

LIST OF CONTRIBUTORS

The numbers in parentheses indicate the pages on which the authors' contributions begin.

GERHARD BAUMANN (77) Reproduction Research Branch, National Institute of Child Health and Human Development, National Institutes of Health, Bethesda, Maryland

JOHAN BOURS (209) Klinisches Institut für Experimentelle Ophthalmologie, Universität Bonn, Bonn-Venusberg, West Germany

NICHOLAS CATSIMPOOLAS (229) Biophysics Laboratory, Department of Nutrition and Food Science, Massachusetts Institute of Technology, Cambridge, Massachusetts

ANDREAS CHRAMBACH (77) Reproduction Research Branch, National Institute of Child Health and Human Development, National Institutes of Health, Bethesda, Maryland

JOHN S. FAWCETT (173) Department of Experimental Biochemistry, London Hospital Medical College, Queen Mary College, London, U.K.

ALEXANDER KOLIN (1) University of California, School of Medicine, Los Angeles, California

BERTOLD J. RADOLA (119) Institut für Lebensmitteltechnologie und Analytische Chemie, Technische Universität München, Freising-Weihenstephan, West Germany

HARRY RILBE (13) Department of Physical Chemistry, Chalmers Institute of Technology, and University of Gothenburg, Gothenburg, Sweden

OLOF VESTERBERG (53) Chemical Division, Department of Occupational Health, National Board of Occupational Safety and Health, Stockholm, Sweden

COLIN W. WRIGLEY (93) Wheat Research Unit, CSIRO, North Ryde, New South Wales, Australia

PREFACE

Isoelectric focusing is a separation method for amphoteric molecules and bioparticles, i.e., species capable of exhibiting positive, negative, or zero electric charge as a function of pH. In the presence of an electric field and a pH gradient, the charged species migrate electrophoretically until they are condensed or "focused" at a position that corresponds to their isoelectric point where their net charge approaches zero. Thus, small molecules, complex macromolecules, and bioparticles can be separated if their isoelectric points (pH of zero charge) are different. This principle has been applied very successfully in the past decade to the separation—at very high resolution—of biological macromolecules, especially proteins. The development of new variations of the technique and its utilization by investigators from a wide variety of disciplines are continuing at a rapid rate.

However, the isoelectric focusing process projects a deceptive simplicity. Despite considerable advancements both in theory and methodology, much remains to be done to improve the technique and to use it in an objectively defined and predictable manner. New approaches to the formation of a stable pH gradient of uniform properties in regard to conductance and viscosity will open up hitherto unrealized possibilities of using the method as a physicochemical tool rather than as a separation technique only. Moreover, the availability of inexpensive means for pH gradient formation will facilitate the use of continuous flow systems which offer promise for the production of relatively large amounts of purified biological material. Improvements are also needed in the development, standardization, and quantitation of two-dimensional maps capable of resolving thousands of macromolecular components simultaneously. Such techniques are at present time-consuming and technically involved, and the results difficult to quantify and interpret. Since their value in future biomedical and biological research is obvious, efforts should be made to provide a rapid and objective means of macromolecular mapping.

This volume is intended to serve in some useful way all those workers who need to know what isoelectric focusing is all about, to aid them in choosing which variation to use in their own research problems, and to point out the areas in which further important contributions can be made. To this end, I was fortunate to gather a group of authors who have been personally involved in molding the isoelectric focusing technique to its present shape. Some of them use a great deal of mathematics, others are concerned with pots, wires, and chemicals. The ultimate goal is the same. As Olof Vesterberg once joked, "if you do not know what to do with a protein, focus it . . . ." It will probably show up as a "heterogeneous" mixture. If you can sustain the first shock, the degradation of your confidence, the guilt of having published results on its "homogeneity," then and only then can the true journey begin in asking and answering what all these bands mean.

1

ISOELECTRIC FOCUSING

**THE PHYSICAL EFFECT, ITS INITIAL DEMONSTRATION,
AND OUTLOOK FOR FUTURE DEVELOPMENTS**

Alexander Kolin

University of California, School of Medicine
Los Angeles, California

I. AN ESSENTIAL CONDITION FOR ISOELECTRIC FOCUSING

In view of the wide importance of isoelectric focusing, reflected in well over 1000 publications in a span of only a few years, it may be of interest to look at the history of the discovery of this phenomenon and to correct some blurred impressions conveyed in the literature.

References in some publications to compartment apparatus (Williams and Waterman, 1929; Tiselius, 1941) as antecedents of the electrophoretic columns in which proteins differing in isoelectric point (pI) by as little as 0.02 pH units can be separated and characterized by their isoelectric points are somewhat misleading. Such instruments consisted typically of numerous compartments (e.g., 14) separated by membranes. The current passing through the electrolyte filling these intruments generated a steady-state pH distribution after periods usually exceeding 24 hr. The pH distribution was a "staircase" with constant pH in each compartment (due to stirring and convection) and a steep jump in pH from compartment to compartment so that the pH gradient was confined to a steep pH jump within the membrane. Such instruments were incapable of providing the essential condition for high-resolution isoelectric analysis based on generation of distinct adjacent zones of ampholytes differing but slightly in their isoelectric pH values. Such analysis presupposes the idea of a *continuous* pH gradient which offers entirely new possibilities. To say that the isoelectric focusing separator is an offspring of compartment apparatus with a staircase pH distribution would be comparable to the assertion that the piano is the ancestor of the violin, an instrument capable of producing a *continuous* sound spectrum like the human voice. As we shall see, such compartment apparatus had no effect on the discovery of the isoelectric focusing effect to which I was led by an entirely different route.

II. ISOELECTRIC FOCUSING—A FAMILY OF THREE EFFECTS

In 1952–1953, I was studying an electrokinetic phenomenon unrelated to electrophoresis: electromagnetophoresis (Kolin, 1953). Its basis is an electromagnetic force, proportional to the vector product of electrical current density and magnetic field intensity, exerted on a particle (which may be electrically neutral) suspended in a conductive fluid whose electrical conductivity differs from that of the particle. In the course of this work I ran into a number of experimental difficulties which included different types of convection. I decided to familiarize myself with the field of electrophoresis in the hope that I might find in the literature ideas which could be applied to the solution of my convection problems.

Dr. H. A. Abramson once gave me a copy of his book "Electrokinetic Phenomena" (Abramson, 1934) which I took as reading matter on a train ride to Fort Wayne, Indiana, where I had to go on a consulting mission. As I was reading about the amphoteric properties of proteins and the existence of an isoelectric point, it became obvious that a titration changing the pH of a solution in time would lead to a concomitant temporal variation in the electrophoretic mobility of a protein ion, reversing its direction of motion at the transition of the pH through the ion's isoelectric point. This image immediately evoked the vision of a "Gedankenexperiment" in which the *temporal* pH variation was replaced by a *space distribution* of pH in an electrophoretic column. I visualized a protein ion dragged through a pH gradient starting with a high positive charge in a low pH region and undergoing a gradual charge diminution on its way to the zone where it would lose its charge and then reverse it on transit through its isoelectric point. Beyond this point the negative charge of the protein ion would grow with the ion's removal from its isoelectric zone in the direction of increasing pH values.

An ion could, of course, be "dragged" by an electric field in such a pH gradient column, but it became clear that such a column would behave very differently from conventional electrophoretic columns in that ions of a given ampholyte would move in opposite directions on either side of their isoelectric zone. It became immediately obvious that there should be two opposite effects:

1.   For a current flowing from low to high pH values, the ions would be swept from all parts of the pH gradient column to converge toward their isoelectric zone where they would lose their charge and come to rest.

2.   For a current flowing in the direction of diminishing pH the positive ampholyte ions on the low pH side of the isoelectric zone as well as the negative ampholyte ions on the high pH side would migrate away from the isoelectric zone, thus evacuating the ampholyte ions from it.

A third aspect of such concentration redistribution of ampholytes in a pH gradient column became clear to me only later when I set about to demonstrate the existence of these predicted effects. It turned out that the ampholyte concentrated in the isoelectric zone exhibited the behavior of an object in stable equilibrium. A displacement of a portion of the zone from the iso-electric region into either a lower or higher pH region (by diffusion or by stirring) would impart to its ampholytes a charge that would automatically return them to the isoelectric zone in the presence of a current flowing in the direction of increasing pH. It was clear that ampholytes of different pI values would form distinct condensation zones at different pH values within the pH gradient column and that such an "isoelectric spectrum" could be used both for preparative separations of proteins, polypeptides, and other ampholytes as well as for their characterization by measuring the pH in the condensation zones (Kolin, 1954, 1955b).

Although I preferred the term isoelectric condensation, I also introduced the term "focusing" in describing the convergence of the ampholyte ions toward their isoelectric zones (Kolin, 1955a, 1958, 1960). This terminology was later adopted by Svensson (Rilbe).

III.   DEMONSTRATION OF THE FOCUSING AND EVACUATION EFFECTS

My first attempts to demonstrate a focusing effect on a protein in a pH gradient were undertaken on strips of filter paper. Hemoglobin was the most convenient protein to use. After initial exploratory experiments, I decided to generate the pH gradient by preparing solutions in which the pH value was linked with a value of density. Thus, stirring the contact region of a dense buffer solution of low pH with an overlying layer of a less dense buffer solution of high pH would result in an intermediate zone of varying density in which the density gradient would be linked with a concomitant pH gradient. The initial experiments were carried out in the simple Lucite apparatus shown in Fig. 1 (Kolin, 1954). L and R are the legs of a U-tube of square cross section (6×6 mm) which communicate with the electrode compartments A and B when the cell is filled with buffer solution. E_1 and E_2 are Pt electrodes that

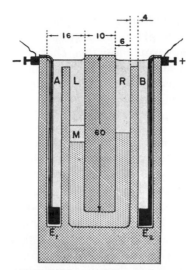

Fig. 1. Electrofocusing cell (dimensions in millimeters). E_1, E_2: electrodes; A, B: electrode compartments; L, R: left and right legs of U-tube; M: middle layer containing protein. [From Kolin (1954).]

are linked to the positive and negative terminals by insulated leads. Liquids of different density are depicted by differences in shading in the figure. The bottom U-shaped fluid body was an acid buffer of pH 2.6 which was made very dense by a high concentration of sucrose. The electrode compartments A and B as well as the upper halves of the legs L and R of the central U-tube were filled with a low-density basic buffer (containing no sucrose) of pH 9.6. In the left leg L, a roughly cubical volume M of a solution of a density intermediate between the densities of the basic and acid buffers was interposed. This M layer contained a dilute solution of hemoglobin; its density was adjusted by addition of sucrose or glycerol. The buffers were prepared according to Michaelis as follows:

(a) *basic buffer:*   4.85 g of sodium acetate + 7.35 g sodium barbital per liter of solution;
(b) *acid buffer:*   prepared by adding to 285 ml of water, 100 ml of the basic buffer + 75 ml of $N/10$ HCl.

Because of turbulence during the process of filling and subsequent diffusion, the M layer is the site of the pH gradient containing a protein solution in a short density gradient column. The electrical current flows from high to low pH in the right column R and from low to high pH in the left leg L as it traverses the M layer. This is the configuration of current and potential gradient required to achieve a condensation of the protein in its isoelectric zone, and the experiment exhibited formation of a sharp zone between the boundaries of the M layer in less than a minute.

The following circumstances accounted for the high speed of the focusing process. The concentration of buffer and protein in the M layer was low. Hence, the M layer was a belt of very low electrical conductivity as compared to the conductivity of the buffer solutions between which it was sandwiched. A large fraction of the potential difference of 220 V between the electrodes was thus applied across the M layer, causing the protein ions in it to move rapidly.

In order to demonstrate simultaneously the predicted focusing effect in a current flowing toward high pH in a pH gradient and evacuation in a current flowing toward decreasing pH values, an experiment was performed in the symmetrical arrangement shown in Fig. 2. Figure 2 shows the U-tube section of the cell of Fig. 1 carrying an M layer in each leg (L and R). Whereas the current flowing toward increasing pH in leg L focuses the hemoglobin within 40 sec in a thin zone between the original boundaries of the left M layer, the same current flowing toward lower pH through the M layer in leg R sweeps the protein toward the boundaries of the M layer where the protein ions slow down and accumulate as they encounter high conductivity buffers at the boundaries of the M layer, where the electrical field intensity (and hence also the ion velocity) is low (Kolin, 1954).

These focusing and evacuation effects can be reversed by reversing the direction of the current. It can also be shown that after disturbing the condensed protein zone in the leg L by a tiny stirrer, and thus rendering it diffuse,

Fig. 2.   Isoelectric focusing and evacuation of hemoglobin in a pH gradient (pH 2.6–9.6) within 40 sec. (a) U-tube with hemoglobin belts in both legs before passage of current. (The polarity is indicated in Fig. 1.) (b) Focusing of hemoglobin by a current flowing in direction of increasing pH in left leg of U-tube. (c) Evacuation of hemoglobin in right leg by a current flowing in direction of decreasing pH. [From Kolin (1954).]

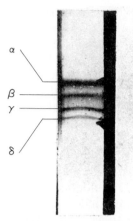

Fig. 3. Isoelectric focusing of a mixture of proteins obtained in 4 min. The top component (α), cytochrome c, is focused nonisoelectrically in a conductivity gradient. The lower ones, hemoglobin (β), catalase (δ), and Azocoll (δ), are focused isoelectrically in pH gradients ranging from pH 4.8 to 7.7. [From Kolin (1955).]

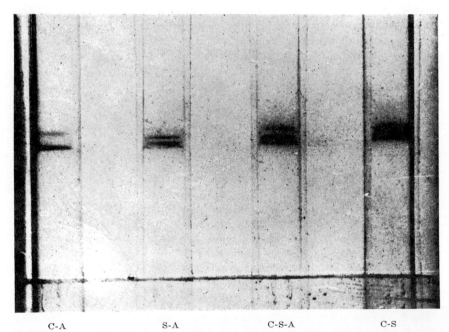

    C-A   S-A   C-S-A   C-S

Fig. 4. Isoelectric fractionation of mixtures of hemoglobins A, C, and S by Tuttle. [From Tuttle: The separation and identification of human hemoglobins by isoelectric line spectra formation, *J. Lab. Clin. Med.* **47**, 811–816, 1956.]

it can be "refocused" by continued passage of current through the M layer, thus demonstrating the stability of the focusing phenomenon.

Figure 3 illustrates the simultaneous focusing of four proteins into narrow zones in an M layer containing a solution of mixed components in a less steep pH gradient ranging in pH from 4.8 to 7.7 (Kolin, 1955b). It must be pointed out that the formation of the top zone (cytochrome c) is not due to isoelectric focusing but to condensation in an electrical conductivity gradient (Kolin, 1955b). In experiments using a somewhat modified apparatus, Tuttle (1956) obtained separation of human hemoglobins in a similar system shown in Fig. 4.

IV.   LIMITATIONS OF THE PREPARED pH GRADIENT METHOD AND THE SUCCESS OF THE NATURAL pH GRADIENT METHOD

The demonstration of isoelectric focusing of proteins in a prepared continuous pH gradient was greeted by Arne Tiselius as an original innovation which seemed to him "to offer great possibilities for further development of electrophoretic methods," as he wrote to me in a letter dated September 6, 1955. He evidently clearly understood the significance of the replacement of the pH staircase distribution, as developed by prolonged electrolysis (basis for natural pH gradient formation) in his own compartment apparatus (Tiselius, 1941), by a continuous pH gradient which is a prerequisite for high-resolution isoelectric separation.

Tiselius' vision was vindicated by the subsequent developments of Svensson (1961, 1962a, b) and Vesterberg (1969) whose analytical success was based on maintenance of a steady-state continuous pH gradient developed by prolonged electrolysis in a column containing a wide spectrum of low-molecular-weight carrier ampholytes synthesized by Vesterberg which reached stationary positions within the column (natural pH gradient). On the other hand, the results obtained initially with the prepared pH gradients (Kolin, 1954, 1955b) were disappointing despite the speed of formation of the isoelectric focusing patterns (Fig. 3). The criticism of the prepared pH gradient method was aptly formulated by Svensson (1960). He pointed out, as did the author (Kolin, 1958, 1960), the deleterious effect of prolonged electric current on the prepared pH gradient, which exhibited a drift under such conditions. The focused protein zones are not localized exactly in the region of their isoelectric pH and their position drifts slowly as they become progressively more diffuse. Isoelectric focusing will be successful only in cases where the drift is not so rapid as to prevent condensation of the ampholytes into isoelectric zones. Ideally, a perfectly stationary pH distribution, unaffected by the current, is required. Svensson reviewed some means of reducing the rate of such drifts, and cited the example of an acetate buffer column with a sodium acetate concentration constant throughout the column. The most acidic portion of the column is also the densest solution because of a high concentration of sucrose and acetic acid, whereas the overlying solution is less concentrated in acetic acid and free of sucrose. A gradient ranging from pH 3 to 6 can thus be obtained which is very stable and accomplishes the desired reduction of drift due to passage of current. Similar buffer columns can be prepared with other weak acids and their salts in analogous fashion.

Further development of isoelectric focusing in prepared pH gradients was inhibited by the spectacular success of Svensson and Vesterberg's focusing technique in natural pH gradients which could be maintained in the presence of current over long periods of time in Ampholine columns (Vesterberg, 1968). In 1958, while visiting Sweden, I met Harry Svensson who became acquainted with the initial experiments on isoelectric focusing in a special seminar held in Tiselius' Institute in 1955. I shared with him my pessimism about the possibility of completely eliminating the drifts in pH gradient columns prepared from conventional buffers. In the course of a long evening walk I mentioned to him an idea, which never progressed beyond a vague qualitative stage, of using mixtures of amino acids and polypeptides of a wide spectrum of isoelectric points instead of conventional buffer mixtures to obtain current-stabilized pH gradient columns. Because of involvement in research on blood flow, I never attempted to follow up this idea which did not approach the sophistication of the final solution achieved by Svensson, to whose thorough insight and experimental skill the method of isoelectric focusing analysis owes its present state of development. By combining the concepts of steady-state pH gradient generation, clearly formulated by Tiselius (1941), with the concept of isoelectric focusing in continuous pH gradient columns (Kolin, 1954, 1955b), and aided by Vesterberg's synthesis of Ampholines (1969), Svensson created a most powerful physicochemical method of protein analysis.

V. SOME RECENT ALTERNATIVE APPROACHES TO GENERATION OF pH GRADIENTS FOR ISOELECTRIC FOCUSING ANALYSIS

Even an excellent method may have some shortcomings which eventually stimulate searches for new approaches to circumvent its limitations. The Svensson–Vesterberg Ampholine column for implementation of the isoelectric focusing principle in a natural pH gradient is no exception, although at present it is by far the best solution.

It is not the intent of this chapter to go into details of methods that are reviewed in subsequent chapters. A mere mention will suffice to provide a summary of alternative approaches serving to circumvent some of the shortcomings of the Ampholine density gradient column technique. Among the drawbacks of this admittedly very effective method is the length of time required to load the column, establish the pH gradient, achieve isoelectric focusing, and finally separate the condensed protein zones. The possibility of interaction between the Ampholines and proteins and the necessity of isolating the separated proteins from the Ampholines should also be mentioned as well as the disturbing ultraviolet absorption of the Ampholine peaks which interferes with spectrophotometric evaluation of the protein fractionation.

A radically different approach to the generation of pH gradients which does not rely on carrier ampholytes or on prolonged electrolysis has been made by taking advantage of the temperature coefficient of the pH of a buffer as well as of the pI of the proteins to be separated (Kolin, 1970; Luner and Kolin, 1970, 1972; Lundahl and Hjertén, 1973). A pH gradient can be established in a buffer solution within seconds by taking advantage of the temperature dependence of the pK. By establishing a temperature gradient within the buffer, pH gradients can be obtained that span a pH range of about 1 pH unit. The slope of the gradient can be controlled simply by varying the temperature limits. This method eliminates the long process of establishing the pH gradient by stationary electrolysis as well as the use of additives such as the Ampholines which have to be separated from the proteins after their isoelectric fractionation.

For example, the pH of a Tris–HCl buffer is determined by the equilibrium relation

$$K = (H^+)(Tris)/(Tris\ H^+) \tag{1}$$

The temperature dependence of the hydrogen ion activity of a buffer solution can be expressed as

$$(\partial p\alpha_H/\partial T) \approx -(\partial \log K/\partial T) - (2z + 1)(\partial \log \gamma/\partial T) \tag{2}$$

where γ is the activity coefficient of an average ion of valence $z = 1$. In this example of the Tris buffer, the second term of Eq. (2) is less than 10% of the first term, and is neglected in the present approximation. The first term can be obtained from the following relation involving the gas constant R and the molar heat of dissociation of an acid $\Delta H°$:

$$(\partial \ln K/\partial T) = (\Delta H°/RT^2) \tag{3}$$

Thus

$$(\partial p\alpha_H/\partial T) \approx -(\Delta H°/RT^2) \tag{4}$$

However, it must be remembered that the temperature affects not only the pH of the buffer but may also affect the pI of the protein to be focused. Focusing can occur only if $(\partial pH/\partial T)_{buf} - (\partial pI/\partial T)_{pro} \neq 0$ (Lundahl and

Hjertén, 1973; Luner and Kolin, 1972). Work on this method has not progressed beyond its initial stages so that it would be premature to assess the promise of this method.

Other interesting possibilities have been described by Lundahl and Hjertén (1973). They include electrophoretically generated pH gradients which are formed by passage of current through solutions containing the gradient of a neutral solute such as sucrose or in acrylamide gels containing a concentration gradient of acrylamide which is used in pore gradient electrophoresis. These authors also reported formation of steep pH gradients at the boundary of buffer solution and acrylamide gel. A particularly promising suggestion of Lundahl and Hjertén (1973) concerns the δ and ε boundaries well known from moving boundary electrophoresis. They point out that these boundaries are *stationary* and that they represent jumps in pH (and conductivity). They suggest the possibility of exploiting these phenomena for generation of stationary pH gradients for isoelectric focusing.

In summary, isoelectric focusing is a physical phenomenon that can be exploited for methodological–analytical purposes in a variety of ways. At the moment the most successful of the techniques of implementation of the principle of isoelectric focusing analysis is the carrier ampholyte column of Svensson and Vesterberg. Some of the new ideas which have evolved during recent years, may, however, lead to alternative solutions of the problem of generating isoelectric spectra of proteins and other ampholytes, which may avoid some of the drawbacks of the currently optimal technique.

REFERENCES

Abramson, H. A. (1934). "Electrokinetic Phenomena." Van Nostrand-Reinhold, Princeton, New Jersey.
Kolin, A. (1953). *Science* **117**, 134–137.
Kolin, A. (1954). *J. Chem. Phys.* **22**, 1628.
Kolin, A. (1955a). *J. Chem. Phys.* **23**, 407.
Kolin, A. (1955b). *Proc. Nat. Acad. Sci. U.S.* **41**, 101.
Kolin, A. (1958). *Methods Biochem. Anal.* **6**, 259–288.
Kolin, A. (1960). *In Medical Physics* (O. Glasser, ed.), Vol. 3, pp. 275–282. Yearbook Publ., Chicago, Illinois.
Kolin, A. (1970). *Methods Med. Res.* **12**, 326–358.
Lundahl, P., and Hjertén, S. (1973). *In* "Isoelectric Focusing and Isotachophoresis" (N. Catsimpoolas, ed.), *Ann. N.Y. Acad. Sci.* **209**, 94–111.
Luner, S. J., and Kolin, A. (1970). *Proc. Nat. Acad. Sci. U.S.* **66**, 898.
Luner, S. J., and Kolin, A. (1972). U.S. Patent 3,664,939.
Svensson, H. (1960). *In* "A Laboratory Manual of Analytical Methods in Protein Chemistry including Polypeptides" (P. Alexander and R. J. Block, eds.), Vol. 1, pp. 195–242. Pergamon, New York.
Svensson, H. (1961). *Acta Chem. Scand.* **15**, 325.
Svensson, H. (1962a). *Acta Chem. Scand.* **16**, 456.

Svensson, H. (1962b). *Arch Biochem. Biophys. Suppl.* **1**, 132.
Tiselius, A. (1941). *Sv. Kem. Tidskr.* **53**, 305.
Tuttle, A. H. (1956). *J. Lab. Clin. Med.* **47**, 811.
Vesterberg, O. (1968). *Sv. Kem. Tidskr.* **80**, 213.
Vesterberg, O. (1969). *Acta Chem. Scand.* **23**, 2653.
Williams, R. R., and Waterman, R. E. (1929). *Proc. Soc. Exp. Biol. Med.* **27**, 56.

2

THEORETICAL ASPECTS OF STEADY-STATE ISOELECTRIC FOCUSING

Harry Rilbe

Department of Physical Chemistry
Chalmers Institute of Technology
and University of Gothenburg
Gothenburg, Sweden

I. INTRODUCTION

Isoelectric focusing in its modern form is closely connected to the concept of carrier ampholytes, which was defined thirteen years ago (Svensson, 1962). The origin of this concept, however, dates back to 1956 in a theoretical paper dealing with the concept of transference numbers of ampholytes (Svensson, 1956). In that article an imaginary experiment with a hypothetical substance was contemplated as follows:

> The relationships which we have been discussing can be still better illustrated by making use of a hypothetical substance containing one acidic and one basic group, the pK values of which are both equal to 7 . . . . We dissolve this substance in pure water; the pH of this solution will of course be 7. At this pH, both polar groups of the ampholyte are half-dissociated, i.e. one fourth of the ampholyte is present in the un-dissociated, one fourth in the zwitterionic, one fourth in the cationic, and one fourth in the anionic form. We can assume that the ampholyte is dissolved in a concentration sufficiently high to allow us to neglect the dissociation of water. Our solution thus has considerable concentrations of ions and must have an appreciable conductance. Both the cations and the anions are, however, subspecies of the same organic radical, being in chemical equilibrium with each other. Consequently, . . . if we carry out a transference experiment with our solution and measure the transport number according to Hartley–Moillet's definition, we must come to the conclusion that the organic radical has a transport number close to 0 because we do not get any electrical transport of the substance. There are equal amounts of cationic and anionic molecular fragments, and since these ions differ by two protons only, they must have nearly the same mobility, and no net transport of the ampholyte can be expected [pp. 33–34].

Although not stated in this text, the pK values mentioned were thought of as intrinsic pK values since pK values directly measurable by titration (representing hybrid dissociation constants) cannot lie closer than 0.6 pH unit (*vide infra*). The assumed degree of ionization ($\alpha = \frac{1}{2}$) for this hypothetical substance checks well with the strict mathematical analysis presented later in this chapter.

It is now well understood that this hypothetical substance was precisely the ideal carrier ampholyte, isoelectric at pH 7. If it were available, it would be able to remove the extremely deep conductivity minimum at pH 7 which had been the most severe obstacle in all earlier attempts at using stationary electrolysis for separation of ampholytes.

This substance is still not available, but in the neutral fraction of Ampholine there may be unknown chemical entities with properties very similar to those of this substance.

The theory of isoelectric focusing has of course been presented before. As a matter of fact, the first paper was a theoretical one (Svensson, 1961), and only recently the author presented a rather exhaustive survey (Rilbe, 1973). In order to avoid undue repetition, aspects which were previously treated rather superficially will be given a more detailed treatment here.

II.   THE DIFFERENTIAL EQUATION
OF ISOELECTRIC FOCUSING

In an electrophoresis column of constant cross-sectional area q (Fig. 1), let the vertical coordinate run in the direction of the current. Let us study the mass flow (mass units per unit area and time) of a protein into and out

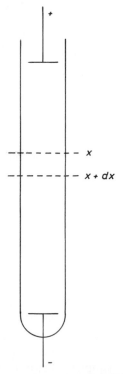

Fig. 1.   Illustration of the theory of electrical and diffusional mass transport in a convection-free column.

of a volume element situated between the levels x and $x + dx$. A positively charged protein migrates into this volume element at the rate

$$J_e = Cui/q\kappa = CuE \tag{1}$$

where C is the protein concentration at the level x, u is its mobility at that point, i is the electric current, κ is the conductivity at the level x, and E is the field strength. The protein also exhibits a diffusional mass flow of magnitude

$$J_d = -D\,(dC/dx) \tag{2}$$

where D is the diffusion coefficient. The minus sign in this equation occurs because the mass flow goes in the direction opposite to that of the concentration gradient dC/dx. The total mass flow relative to the chosen frame of reference (e.g., the column) is thus

$$J = CuE - D\,(dC/dx) \tag{3}$$

At the level $x + dx$ the corresponding mass flow is

$$J + dJ = CuE - D\frac{dC}{dx} + \frac{d}{dx}\left[CuE - D\frac{dC}{dx}\right]dx \tag{4}$$

The mass increase per unit area and time thus becomes

$$-dJ = \frac{d}{dx}\left[-CuE + D\frac{dC}{dx}\right]dx \tag{5}$$

Multiplication by q gives the mass increase per unit time, and subsequent division by the volume $q\,dx$ gives the concentration increase per unit time. With two independent variables, level x and time t, we have to use partial derivatives:

$$\frac{\partial C}{\partial t} = -\frac{\partial(CuE)}{\partial x} + \frac{\partial}{\partial x}\left[D\frac{\partial C}{\partial x}\right] \tag{6}$$

This is the general differential equation of simultaneous electrophoretical and diffusional mass transport.

   If the protein is negatively charged, it exhibits an inflow through the level $x + dx$ and an outflow through the level x. This does not require a new set of equations since it is taken care of by the negative sign of the mobility u.

   In a steady state, as in ideal isoelectric focusing, everything is constant in time, which yields the much simpler differential equation

$$\frac{d(CuE)}{dx} = \frac{d}{dx}\left[D\frac{dC}{dx}\right] \tag{7}$$

which on integration gives

$$CuE = D \frac{dC}{dx} + \text{integration constant} \tag{8}$$

This equation must be valid everywhere in the column, even outside the focused protein zone, where the protein under consideration is absent. It follows that the integration constant is zero, and we obtain

$$CuE = D \, (dC/dx) \tag{9}$$

This is the general differential equation for the final steady state in isoelectric focusing. It should be remembered that it is not applicable to components which cannot reach a steady state. It is thus completely inapplicable to nonelectrolytes that do not form complexes with electrolytes since $u = 0$ requires a vanishing concentration gradient. Any varying concentration of such a nonelectrolyte thus violates Eq. (9). However, a constant concentration gradient of sucrose, much used as an anticonvective measure in isoelectric focusing in columns, does not violate Eq. (8) if it can be kept reasonably constant in time also.

Even electrolytes may fail to come to a steady state. For instance, when one strong and one weak acid simultaneously migrate to the anode, the strong one collects there more rapidly than the weak one since the former suppresses the dissociation of the latter. The migration of the weak acid is then stopped, but it continues to diffuse through the strong acid layer to the anode. Since no electric mass transport can balance this diffusional mass transport, a steady state cannot be reached.

Components unable to reach a steady state may slowly alter the conditions underlying the steady state of other components. The latter states then become only quasi stationary, subject to a slow drift.

Equation (9) was first derived for a protein mainly because the analysis and separation of proteins comprise the main purpose of the method, but also because both mobilities and diffusion coefficients of proteins are well-defined and measurable quantities. In the isoelectric state the mobility is zero, and the diffusion coefficient can be measured without difficulty. On either side of the isoelectric point the protein carries an electric charge, which implies a certain complication in a diffusion measurement, but the diffusion coefficient can still be measured in the presence of a sufficient amount of salt, and it is found to be essentially independent of pH. The validity and applicability of Eq. (9) for chemical entities other than proteins will now be elucidated.

Equation (9) cannot be applied to arbitrary electrolytes. Consider for instance potassium sulfate (K_2SO_4). This salt has a well-defined and measurable diffusion coefficient, but it has two mobilities, one positive for K^+ and one negative for SO_4^{2-}. The equation thus becomes meaningless for a salt. As far as isoelectric focusing is concerned, there is also no reason to try to use it for a salt since all salts are split up in acid and base collecting at anode and cathode, respectively.

Since mobilities can be ascribed to ion constituents only, we could argue that Eq. (9) has to be applied to ion constituents. However, the difficulty then arises that diffusion coefficients of ions cannot be measured, and the equation again becomes meaningless. It is true that so-called self-diffusion coefficients can be measured by the use of radioactive isotopes, but this kind of coefficient can hardly be accepted in a theory for isoelectric focusing. In a self-diffusion experiment an exchange occurs between radioactive and inactive ions of the same kind, whereby the diffusion can take place without violation of the electroneutrality condition and without simultaneous diffusion of another ion. This is not the mechanism of the diffusion occurring in isoelectrically focused zones. The diffusion coefficient in Eq. (9) thus has to be ascribed to a chemical substance, whereas the mobility must be ascribed to an ion constituent. This is the logical dilemma that has to be solved before we can proceed to an interpretation of the equation.

Isoelectric focusing is utilized in order to separate protolytes from each other. Acids and bases migrate to the anode and cathode, respectively, and are concentrated there, whereas ampholytes condense in isoelectric zones in the order of their isoelectric points. For this reason it is adequate to ascribe the diffusion coefficient in Eq. (9) to protolytes. Strong bases, which are not protolytes in the Brønsted sense, have to be included since they occur at the cathode in isoelectric focusing. In the pre-Brønsted terminology, diffusion coefficients of acids, bases, and ampholytes will have to be used in Eq. (9).

A protolyte, like all other binary electrolytes, has two ion constituents, the cationic being H^+, whereas strong bases have OH^- as the anionic constituent. Since the solvent ion concentrations are unambiguously defined by the concentrations of all protolytes and strong bases, there is no reason to apply Eq. (9) to the solvent ions H^+ and OH^-. The mobilities to be inserted in this equation are consequently the mobilities of those ion constituents which are not H^+ or OH^-. Hence, for sulfuric acid, Eq. (9) should be satisfied by insertion of the mobility of the SO_4^{2-} ion constituent and the diffusion coefficient of H_2SO_4; for acetic acid, by insertion of the mobility of the Ac^- ion constituent (which approximates zero in strongly acidic media because of the suppressed dissociation of the acid) and the diffusion coefficient of HAc; for sodium hydroxide, by insertion of the mobility of Na^+ and the diffusion coefficient of NaOH. For ampholytes such as amino acids, peptides, and proteins, it is superfluous to state that the mobility refers to the anion constituent and not to H^+, since it is self-evident that we mean the ion which really contains the organic chemical structure. It should be understood, however, that there are only purely semantic reasons why the mobility of a protein is a meaningful concept whereas the mobility of sulfuric acid is not.

III. CONVECTION-FREE ELECTROLYSIS OF A SINGLE STRONG ACID OR STRONG BASE

A. Concentration Course of Acid or Base for Constant Transference Numbers

Elimination of E and u from Eq. (9) with the aid of the relations

$$E = i/q\kappa \tag{10}$$

$$t\kappa = FCzu \tag{11}$$

where F is Faraday's constant and t is the transference number of the ion constituent with mobility u, gives the differential equation in the form

$$(i/Fqz)\,dx = (D/t)\,dC \tag{12}$$

The factor in front of dx is constant whereas both D and t on the right-hand side vary with C. If the functions $D(C)$ and $t(C)$ are known from measurements, it is possible to integrate the equation and thus to calculate C as a function of x.

For strong electrolytes D and t vary only moderately with C. An assumption of constant diffusion coefficients and transference numbers will thus give an approximately correct picture of the concentration courses of strong acids and bases during stationary electrolysis. This assumption leads to a linear concentration course which can be described by the equation

$$C = ti(x - b)/FDzq \tag{13}$$

where the distance b serves as an integration constant. Since negative concentrations cannot be allowed, the following restriction imposes the integration constant:

$$(x - b)/z \geq 0 \tag{14}$$

for all x values between the electrodes. If the electrode coordinates are defined as $x = -a/2$ (anode) and $x = +a/2$ (cathode), condition (14) implies for a strong base ($z > 0$)

$$b \leq -a/2 \tag{15}$$

and for a strong acid ($z < 0$)

$$b \geq +a/2 \tag{16}$$

Alternatively, we can state that Eq. (13) is valid within the x interval, making C positive or zero, and that the concentration is zero in the remainder of the electrolyzer. This also satisfies Eq. (9), which has a trivial solution $C = 0$. The experimental possibility of forcing an electric current through the pure solvent is a question of its conductivity and of cooling facilities. In the following treatment, the restriction (14) will be assumed to hold.

The linear concentration course can also be expressed using an integration constant C_0 with the physical meaning of the initial constant concentration before electrolysis is started; this concentration remains unchanged in the center of the apparatus ($x = 0$):

$$C = C_0 + tix/FDzq \tag{17}$$

The mass content of the apparatus is then $C_0 aq$ mass units. The concentration gradient

$$dC/dx = ti/FDzq \tag{18}$$

is recognized to be proportional to the current density and to the transference number of that ion constituent which is not H^+ or OH^-. It is inversely proportional to the diffusion coefficient of the acid or base.

For a certain critical current i_c the concentration vanishes at one of the electrodes. Equation (17) shows that this occurs for the current

$$i_c = \frac{2FDC_0|z|q}{ta} \tag{19}$$

With the actual current i expressed as a certain fraction k of this limiting current,

$$i = ki_c \qquad (0 \leq k \leq 1) \tag{20}$$

Eq. (17) takes the form

$$C = C_0(1 \pm 2kx/a) \tag{21}$$

where the upper sign is to be taken for bases, the lower sign for acids.

B. The Ohmic Resistance of the Electrolyzer

The electric resistance of the electrolyzer is now considered neglecting the conductivity of the pure solvent. The conductivity of a solution is then given by the equation

$$\kappa = F\Sigma C_i z_i u_i \tag{22}$$

where the C_i are the individual molar concentrations of the ion constituents, the z_i are their valencies (with signs), and the u_i are their mobilities (with signs).

For a strong acid or a strong base, only two ion constituents are present, and because of the electroneutrality condition Cz has the same numerical value and can be put outside the summation sign:

$$\kappa = FC|z|(u_+ - u_-) \tag{23}$$

where C and z refer to that ion constituent which is not H^+ or OH^-, and u_+ is the cationic and u_- the (negative) anionic mobility.

Before the current is started, the concentration C_0 prevails in the whole apparatus, and its resistance R_0 becomes

$$R_0 = \frac{a}{q\kappa} = \frac{a}{FC_0|z|q(u_+ - u_-)} \tag{24}$$

When current flows, the conductivity rises in one end and falls in the other. For calculation of the resistance, we have to integrate the differential equation

$$dR = dx/q\kappa \tag{25}$$

where κ from (23) and C from (21) are to be inserted. Finally, division by (24) gives the result

$$R = R_0 \frac{1}{2k} \ln \frac{1 + k}{1 - k} \tag{26}$$

for both acid and base. The ratio R/R_0 as a function of k is given in Fig. 2. The resistance rises to infinity very rapidly as k approaches unity. For $k = 0.95$, the resistance is still less than twice the initial resistance. The rise to infinity for $k = 1$ is of course because the conductivity of the pure solvent has not been considered.

If the current density is allowed to rise beyond the critical value given by Eq. (19), a layer of pure solvent will develop close to one of the electrodes, and the point $x = b$ in Eq. (13) then lies within the electrolyzer. The conductivity of the pure solvent then has to be taken into account. Since this is very low, 5.5×10^{-8}/ohm cm for water at 25°C and 1.86×10^{-8}/ohm cm at 0°C, it is necessary to be familiar with field strengths and heat effects in pure water at various current densities. Table I gives some information on this point. As is evident from this table, current densities higher than about 10^{-4} A/cm^2 in pure water are unrealistic with reference to the voltages generally available and also with regard to the efficiency of easily designed cooling systems.

C.   Comparison with Tiselius' Experiments

A qualitative verification of Eq. (13) is available in Tiselius' (1941) paper Stationary Electrolysis of Ampholyte Solutions. He obtained the rectilinear concentration courses shown in Fig. 3 on electrolysis of dilute solutions of sodium hydroxide and sulfuric acid in an electrolyzer with 12 compartments

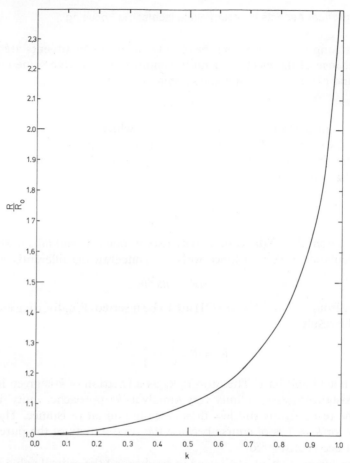

Fig. 2. The ohmic resistance of an electrolyzer charged with a single strong acid or base as a function of the electric current in relation to the critical current (19) for which the concentration drops to zero at one of the electrodes.

TABLE I

Field Strength ($E = i/q\kappa$) and
Heat Effect ($P = i^2/q^2\kappa$) in
Electrolyte-Free Water at 25°C
and Various Current Densities i/q

i/q cm^2/A	E cm/V	P cm^3/W
0.00001	182	0.00182
0.00003	545	0.0164
0.0001	1818	0.182
0.0003	5450	1.64
0.001	18182	18.2

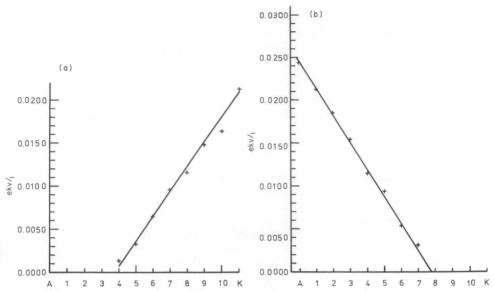

Fig. 3.   Results of Tiselius' (1941) measurements on electrolysis of strong acids and bases. (a) Stationary electrolysis of NaOH at a current of 5.7 mA. (b) Stationary electrolysis of H_2SO_4 at a current of 8 mA.

denoted by A (anode compartment), the numbers 1–10, and K (cathode compartment). As will be shown, the data in Fig. 3, together with additional data in Tiselius' text, also suffice to render a quantitative verification of the theory presented here, including the interpretation of the mobilities as pertaining to the ion constituents Na^+ and SO_4^{2-}, and the diffusion coefficients as pertaining to the substances NaOH and H_2SO_4.

With cell number n as the independent variable, the straight lines in Fig. 3 are represented by the equations

$$C = 2.86(n - 3.8) \quad \text{moles/m}^3 \qquad \text{(NaOH)} \qquad (27)$$

$$C = -1.55(n - 7.9) \quad \text{moles/m}^3 \qquad (\text{H}_2\text{SO}_4) \qquad (28)$$

Thus the concentration gradients are

$$dC/dn = 2.86 \quad \text{moles/m}^3 \qquad \text{(NaOH)} \qquad (29)$$

$$dC/dn = -1.55 \quad \text{moles/m}^3 \qquad (\text{H}_2\text{SO}_4) \qquad (30)$$

In order to compare these experimental data with the theoretical prediction in Eq. (18), numerical values of the transference numbers of the ion constituents and of the diffusion coefficients of the substances are required.

Since low concentrations of the two solutes were used, equivalent conductivities pertaining to infinite dilution, available in Conway's (1952) book, can be inserted in Eq. (18). The diffusion coefficients at infinite dilution can then be calculated by using the Nernst–Hartley equation

$$D^0 = \frac{RT(z_+ - z_-)t_+{}^0 t_-{}^0 \Lambda^0}{-F^2 z_+ z_-} \tag{31}$$

(see Robinson and Stokes, 1959, p. 288). The equivalent conductivities, transference numbers, and diffusion coefficients thus found are collected in Table II.

TABLE II

Equivalent Conductivities (λ), Transference Numbers (t), and Diffusion Coefficients (D) at Infinite Dilution and 25°C

Ion constituent; substance	λ ohm equiv/cm^2	t	D 10^5 sec/cm^2
H^+	349.8	0.8142	—
OH^-	197.6	0.7977	—
Na^+	50.11	0.2023	—
SO_4^{2-}	79.81	0.1858	—
NaOH	247.71	—	2.127
H_2SO_4	429.61	—	2.594

The currents in Tiselius' two experiments were 5.7 mA for electrolysis of NaOH and 8 mA for electrolysis of H_2SO_4. With the data for t and D taken from Table II, and with $z = 1$ for NaOH and $z = -2$ for H_2SO_4, we thus obtain the following theoretically predicted concentration gradients:

$$ti/FDzq = 5.62/q \quad \text{moles/m}^2 \qquad \text{(NaOH)} \tag{32}$$

$$ti/FDzq = -2.97/q \quad \text{moles/m}^2 \qquad \text{(H}_2\text{SO}_4) \tag{33}$$

These data have to be compared with the experimental values given in (29) and (30) after division by the effective cell thickness d in the current direction. The ratio between experimental and theoretical concentration gradients becomes

$$\frac{dC}{dx}\frac{FDzq}{ti} = \frac{0.509q}{d} \quad \text{m}^{-1} \qquad \text{(NaOH)}$$

$$\frac{dC}{dx}\frac{FDzq}{ti} = \frac{0.522q}{d} \quad \text{m}^{-1} \qquad \text{(H}_2\text{SO}_4)$$

Because of the close agreement these equations can be united into one:

$$\frac{dC}{dx}\frac{FDzq}{ti} = \frac{0.515q}{d} \quad (1 \pm 0.014) \quad \text{m}^{-1} \tag{34}$$

The author takes the small discrepancy of only $\pm 1.4\%$ as evidence for the correctness of Eqs. (13), (17), and (18) and of the interpretation of u and D in Eq. (9), as described in Section II.

For full quantitative verification we should also require q/d to be about 1.94 m ($= m/0.515$). The information available on this point in Tiselius' article consists of the cell dimensions given as $40 \times 40 \times 5$ mm^3. These figures, taken as they stand, would indicate $q/d = 1600/5 = 320$ mm, a distance that is six times too small. In addition, the cell dimensions cannot be used directly because Tiselius had active stirring (a system of vertical oscillating glass rods) in the compartments and therefore he cannot possibly have had liquid up to the upper edges of the paper membranes. With a safety margin of 10 mm, the cross-sectional area is reduced to 12 cm^2 and the q/d length to 0.24 m, which is about eight times smaller than 1.94 m.

This large discrepancy occurs because the separation process takes place exclusively within the membranes, whereas the free solutions between them are inactive in this respect. This reduces the cross-sectional area still more to that part of the membrane which is available to ion migration, but it also reduces the effective cell thickness d to the effective path length across a membrane. Tiselius' article does not contain enough detail about the membranes for calculation of their q and d characteristics. The only information given on this point is the phrase "The membranes used were mostly of hardened filter paper."

It is impossible today to figure out what brand of filter paper Tiselius used, but the author has made some simple measurements on Munktell's Genuine Swedish Filtering Paper No. 00H, which is described as a rather thick, very hard paper. Such papers were weighed both dry and moistened with water, and the mass fraction of cellulose in the moistened paper was thus found to be 0.407. Since the density of cellulose is about 1.6 g/cm^3, its volume fraction becomes 0.301. This implies that the fibers occupy the fraction $(0.301)^{1/3} = 0.67$ of the path length along a straight line through the paper, and that they occupy the fraction $(0.301)^{2/3} = 0.45$ of the area of a plane through the paper. One thus arrives at a probable cross-sectional area of $12 \times 0.55 = 6.6$ cm^2. The requirement $q/d = 194$ cm then gives a thickness $d = 0.34$ mm. The thickness of the paper in the dry condition was 0.20 mm. With due regard to the swelling of cellulose fibers in water and to the fact that 0.67 of the paper thickness is occupied by cellulose fibers, an effective migration path length of 0.34 mm does not seem unreasonable, especially as a tortuous path of ions within the membrane also implies a reduced field strength along such a path (cf. Kunkel and Tiselius, 1951).

As is evident from Fig. 3, Tiselius used current densities greatly in excess of the critical one given by Eq. (19). With $i = 8$ mA, as in his H$_2$SO$_4$ experiment, and with the probable cross-sectional area deduced above, we get a current density of 0.0012 A/cm^2 within the paper membranes. With electro-lyte-free water this corresponds to a field strength of 21,800 V/cm and a heat effect of 26 W/cm^3 (cf. Table I). Since no cooling facilities were described, these figures are quite unrealistic. The only possible explanation is that the acid- and alkali-free compartments were open to the atmosphere and thus could collect carbon dioxide as well as dust. In a closed electrolyzer such absorption of electrolytes from the surroundings cannot occur. The critical current given by Eq. (19) then assumes the character of a limiting current that can hardly be surpassed.

On the basis of Eqs. (27) and (28) and with the aid of activity coefficients available in Conway's (1952) book, the pH courses presented in Fig. 4 can be calculated for Tiselius' experiments. They are quite useful in the pH ranges 1.7–2.6 and 11.5–12.3.

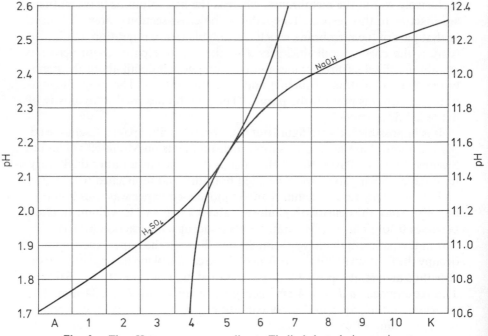

Fig. 4.   The pH courses corresponding to Tiselius' electrolysis experiments.

D. Convection-Free Electrolysis of a Salt

When a salt is electrolyzed under convection-free conditions, acid and base are completely separated, and consequently equations already derived apply to each of them. Both acid and base acquire concentration courses characterized by a linear decline from the respective electrode toward the middle of the apparatus. Because of the very high resistance of pure solvent, we can regard a steady state in which acid and base get vanishing concentrations at the same point, say $x = x_0$, as a limiting degree of separation that can hardly be surpassed. The point at which this occurs depends on the transference numbers and diffusion coefficients of acid and base as shown by the following analysis.

According to Eq. (13), the equivalent concentration of acid or base is

$$\pm Cz = ti(x - x_0)/FDq \tag{35}$$

where the minus sign is to be used for the acid. Multiplication by q and integration from $-a/2$ to x_0 for the acid and from x_0 to $+a/2$ for the base gives the total amounts of acid and base, which are equal. This gives the equation

$$(t/D)_a(a/2 + x_0)^2 = (t/D)_b(a/2 - x_0)^2 \tag{36}$$

where subscript a means acid and subscript b means base. This can be solved for x_0 with the result

$$\frac{2x_0}{a} = \frac{[(t/D)_b]^{1/2} - [(t/D)_a]^{1/2}}{[(t/D)_b]^{1/2} + [(t/D)_a]^{1/2}} \tag{37}$$

Because of the square-root dependence, this point of neutrality always lies fairly close to the center of the electrolyzer. For sodium sulfate it lies only $0.035a$ from the center (toward the cathode) in spite of the large differences in both t and D (in opposite directions) between sodium hydroxide and sulfuric acid (cf. Table II).

IV. CONVECTION-FREE ELECTROLYSIS OF A SINGLE WEAK ACID OR WEAK BASE

A. Basic Equations

Electrolysis of weak acids and bases rather than of strong ones is no reason for any material change in the electromigration theory. The transference numbers of the ions of weak acids and bases are also essentially independent of concentration. Since most of the acid or base is present in an undissociated form, the diffusion coefficient cannot be calculated by Eq. (31), which is valid for strong electrolytes only, but is open to measurement. Consequently, approximately linear concentration courses should also be expected on electrolysis of weak acids and bases, but of course the pH gradient becomes different. For calculation of the latter, the dependence of pH on the concentration of a weak acid or base must be elucidated.

With activities of solutes denoted by [ ] and concentrations by ( ), the mass action law, the ion product of water, and the electroneutrality condition give the following three equations for a solution of a weak acid:

$$[H^+][A^-] = K[HA] \tag{38}$$

$$[H^+][OH^-] = K_w^2 \tag{39}$$

$$(H^+) = (OH^-) + (A^-) \tag{40}$$

For convenience, the hydrogen ion activity will be denoted by h instead of $[H^+]$, and with introduction of f for the activity coefficient of a monovalent ion and of α for the degree of dissociation, Eqs. (39) and (40) can be written

$$hf(OH^-) = K_w^2 \tag{41}$$

$$h/f = (OH^-) + C\alpha \tag{42}$$

where C is the total concentration of the acid. Elimination of (OH^-) gives

$$hf(h/f - C\alpha) = K_w^2 \tag{43}$$

In Eq. (38) the activity of HA will be assumed equal to the concentration, and on introduction of the degree of dissociation we obtain

$$hf\alpha = K(1 - \alpha) \tag{44}$$

Elimination of α between (43) and (44) finally gives

$$C = (h^2/K + h/f)(1 - K_w^2/h^2) \tag{45}$$

This equation, which is of quite general validity, relates h, and thus pH, with the concentration of the acid. Since it is of the third degree in h, the pH of a solution of given concentration cannot readily be calculated, but for a given pH the corresponding concentration is easily calculated for $f = 1$, and the activity coefficient can be obtained with the aid of the Debye–Hückel theory.

For a weak base we have instead of Eq. (38),

$$[H^+][A] = K[HA^+] \tag{46}$$

and instead of (40),

$$(H^+) + (HA^+) = (OH^-) \tag{47}$$

The procedure for solution is the same, and it leads to the following equation, which substitutes Eq. (45):

$$C = (K_w^2 - h^2)(1/fh + K/h^2) \tag{48}$$

B. Calculation of the pH Gradient

With the definitions

$$c = \text{the unit concentration of 1 mole/liter} \tag{49}$$

$$pH = -\log h/c \tag{50}$$

$$pK = -\log K/c \tag{51}$$

we gain the great advantage of dimensionally homogeneous equations in all instances and of avoiding meaningless expressions such as $\log K$ and $\log h$. These definitions also allow arbitrary concentration units for h and K, but the logarithms cannot be taken until they have been converted to the unit of c, or vice versa.

Differentiation of (50) now gives

$$dh/d(\text{pH}) = -h \ln 10 \tag{52}$$

and differentiation of (45) yields

$$\frac{dC}{dh} = \frac{2h}{K} + \frac{1}{f}(1 + K_w^2/h^2) \tag{53}$$

Multiplication of (52) and (53) and inversion gives the pH dependence on concentration:

$$\frac{d(\text{pH})}{d(C/c)} = \frac{-\log e}{(2h^2/Kc) + (h/fc)(1 + K_w^2/h^2)} \tag{54}$$

Multiplication with the concentration gradient dC/dx in the electrolyzer finally gives the pH gradient in the latter. This is superfluous, however, since we have reason to expect a constant concentration gradient in the apparatus. We can just as well analyze Eq. (54).

At pH 7, $h = K_w$, we obtain

$$\frac{d(\text{pH})}{d(C/c)} = \frac{-Kc \log e}{2K_w(K_w + K/f)} \qquad (\text{pH} = 7) \tag{55}$$

This is a very large number unless the acid is extremely weak. The neutral range is useless also for another reason: the conductivity there is so low that it tends to quench the current.

Ignoring the neutral range and going down to pH 6 and lower, we observe that the term K_w^2/h^2 becomes negligibly small and that Eq. (54) simplifies to

$$\frac{d(\text{pH})}{d(C/c)} = \frac{-\log e}{2h^2/Kc + h/fc} \tag{56}$$

For weak bases and pH > 8, the corresponding equation is

$$\frac{d(\text{pH})}{d(C/c)} = \frac{fch^2 \log e}{K_w^2(2Kf + h)} \tag{57}$$

1. Most Acidic (Most Alkaline) Range

If the first term in the denominators of Eqs. (56) and (57) dominates over the second term, and it does so for

$$\text{pH} < p(K/f) - 0.7 \qquad (\text{weak acids}) \tag{58}$$

$$\text{pH} > p(Kf) + 0.7 \qquad (\text{weak bases}) \tag{59}$$

then the pH gradient is given by the equations

$$\frac{-d(\text{pH})}{d(C/c)} = \frac{Kc \log e}{2h^2} \qquad \text{(weak acids)} \qquad (60)$$

$$\frac{d(\text{pH})}{d(C/c)} = \frac{ch^2 \log e}{2KK_w^2} \qquad \text{(weak bases)} \qquad (61)$$

Since these expressions are positive and dimensionless, the logarithms can be taken:

$$\log \frac{-d(\text{pH})}{d(C/c)} = 2\text{pH} - \text{p}K - 0.66 \qquad \text{(weak acids)} \qquad (62)$$

$$\log \frac{d(\text{pH})}{d(C/c)} = -2\text{pH} + \text{p}K + 13.34 \qquad \text{(weak bases)} \qquad (63)$$

These logarithms vanish for a pH change of 1 unit per concentration change of 1 mole/liter, which represents a very useful pH gradient. This occurs for

$$\text{pH} = \tfrac{1}{2}\text{p}K + 0.33 \qquad \text{(weak acids)} \qquad (64)$$

$$\text{pH} = \tfrac{1}{2}\text{p}K + 6.67 \qquad \text{(weak bases)} \qquad (65)$$

At a pH half a unit closer to neutrality, a 10 times steeper pH course prevails, a trend that persists to the limits (58) and (59).

2. Medium Acidic (Medium Alkaline) Range

If both terms in the denominators of Eqs. (56) and (57) are of the same order of magnitude, which occurs for

$$\text{pH} \approx \text{p}(K/f) + 0.30 \qquad \text{(weak acids)} \qquad (66)$$

$$\text{pH} \approx \text{p}(Kf) - 0.30 \qquad \text{(weak bases)} \qquad (67)$$

the pH gradient becomes

$$\frac{-d(\text{pH})}{d(C/c)} = \frac{f^2 \log e}{K/c} \qquad \text{(weak acids)} \qquad (68)$$

$$\frac{d(\text{pH})}{d(C/c)} = \frac{f^2 Kc \log e}{K_w^2} \qquad \text{(weak bases)} \qquad (69)$$

For acetic acid with a pK of about 4.75, this corresponds to a pH shift of 0.1 unit for a concentration change of 4 μmoles/liter. Such a pH course is far too steep to be useful.

3.   Least Acidic (Least Alkaline) Range

If the last term in the denominators of Eqs. (56) and (57) dominates, which occurs for

$$pH > p(K/f) + 1.30 \qquad \text{(weak acids)} \tag{70}$$

$$pH < p(Kf) - 1.30 \qquad \text{(weak bases)} \tag{71}$$

the pH gradient becomes still stronger than that in Eqs. (68) and (69). Such gradients are not useful for the present purpose.

C.   Calculation of the Activity Coefficient

According to Güntelberg (1926), the activity factor for a monovalent ion, as derived by the Debye–Hückel theory, can be written in the form

$$\log f = -0.5092\sqrt{I}/(\sqrt{c} + \sqrt{I}) \tag{72}$$

for water solutions at 25°C, where I is the ionic strength expressed in moles as mass unit and liter as volume unit. In the case of a single weak acid, the ionic strength is identical to the ion concentration, which is h/f:

$$I = h/f \tag{73}$$

Insertion of this and some manipulation with the logarithm gives the equation

$$\ln \frac{1}{f^{1/2}} = \frac{0.5862(h/c)^{1/2}}{f^{1/2} + (h/c)^{1/2}} \tag{74}$$

By passing over to the exponential function on both sides and by performing a series expansion with three terms, we obtain

$$\frac{1}{f^{1/2}} = 1 + \frac{0.5862(h/c)^{1/2}}{f^{1/2} + (h/c)^{1/2}} + \frac{(0.5862)^2(h/c)}{2[f^{1/2} + (h/c)^{1/2}]^2} \tag{75}$$

The last term is smaller than 0.01 under the condition

$$h/f < 0.10 \ c, \qquad -\log(H^+)/c > 1 \tag{76}$$

and will be neglected. With sufficient accuracy we have

$$\frac{1}{f^{1/2}} = 1 + \frac{0.5862(h/c)^{1/2}}{f^{1/2} + (h/c)^{1/2}} \tag{77}$$

This equation is of the second degree with reference to $f^{1/2}$. Since the solution of a second-degree equation is inconvenient, it is preferable to write the first approximation of the solution as

$$f^{1/2} = 1 - a(h/c)^{1/2} \tag{78}$$

with an unknown coefficient a. To determine a, Eq. (78) is inserted in (77), and in the resulting equation the coefficient of $(h/c)^{1/2}$ is put equal to zero. This gives $a = 0.5862$ and

$$f^{1/2} = 1 - 0.586(h/c)^{1/2} \tag{79}$$

$$1/f = 1 + 1.17(h/c)^{1/2} \tag{80}$$

This result can be introduced into the preceding equations. The activity factor is 0.99 or closer to unity for pH values between 4.14 and 9.86 if a weak acid or base is the only solute present. Numerical values for pH values between 4.1 and 2.0 are given in Table III.

TABLE III

The Activity Factor as a
Function of pH

pH	$1/f$	pH	$1/f$
4.1	1.010	3.0	1.037
4.0	1.012	2.9	1.042
3.9	1.013	2.8	1.047
3.8	1.015	2.7	1.053
3.7	1.017	2.6	1.059
3.6	1.019	2.5	1.066
3.5	1.021	2.4	1.074
3.4	1.023	2.3	1.083
3.3	1.026	2.2	1.093
3.2	1.030	2.1	1.105
3.1	1.033	2.0	1.117

D. Calculation of pH as an Explicit Function of Concentration

We have found that a useful pH gradient in a solution of a weak acid is to be found only in the pH range below pK. In this region Eq. (45) can be simplified to

$$h^2 + (K/f)h = KC \tag{81}$$

Since the second term is smaller than the first, a first approximation of its solution can be written as

$$h = (KC)^{1/2}, \qquad \text{pH} = \tfrac{1}{2}(\text{p}K + \text{p}C) \tag{82}$$

where pC is defined as

$$\text{p}C = -\log C/c \tag{83}$$

The exact solution of (81) is

$$h = -K/2f \pm [(K/2f)^2 + KC]^{1/2} \tag{84}$$

The minus sign has to be discarded, and in order to bring the solution into a close relation to the approximate solution (82), it is written in the form

$$h = (KC)^{1/2}(1 + K/4f^2C)^{1/2} - K/2f \tag{85}$$

By using the binomial series including the first three terms, we obtain

$$h = (KC)^{1/2}\left[1 - \frac{(K/C)^{1/2}}{2f} + \frac{K/C}{8f^2} - \frac{(K/C)^2}{128f^4}\right] \tag{86}$$

If

$$f^2C > 0.884K \tag{87}$$

this solution can be abbreviated to

$$h = (KC)^{1/2}\left[1 - \frac{(K/C)^{1/2}}{2f} + \frac{K/C}{8f^2}\right] \tag{88}$$

By taking the logarithms of both members and by using (50), (51), and (83), we arrive at the equation

$$\text{pH} = \tfrac{1}{2}(\text{p}K + \text{p}C) - \log\left[1 - \frac{(K/C)^{1/2}}{2f} + \frac{K/C}{8f^2}\right]$$

$$= \tfrac{1}{2}(\text{p}K + \text{p}C) + \Delta\text{pH} \tag{89}$$

The correction term ΔpH thus defined can be simplified by conversion to a natural logarithm and by series expansion with inclusion of two terms as follows:

$$\Delta\text{pH} = -\log e \ln\left[1 - \frac{(K/C)^{1/2}}{2f} + \frac{K/C}{8f^2}\right]$$

$$= -\log e\left[\left(-\frac{(K/C)^{1/2}}{2f} + \frac{K/C}{8f^2}\right) - \frac{1}{2}\left(-\frac{(K/C)^{1/2}}{2f} + \frac{K/C}{8f^2}\right)^2\right]$$

$$= -\log e\left[-\frac{(K/C)^{1/2}}{2f}\right] = \frac{(K/C)^{1/2}}{2f}\log e \tag{90}$$

Powers of K/C higher than unity have to be discarded in this development since the approximation (88) ends with this order of magnitude. We thus obtain the final equation

$$\text{pH} = \tfrac{1}{2}(\text{p}K + \text{p}C) + \frac{(K/C)^{1/2}}{2f}\log e \tag{91}$$

The activity factor in this equation has the character of a correction factor within a correction term, and therefore its influence is very small. By using Eqs. (82) and (80) for the correction within the correction, it is readily proved that f in Eq. (91) can be omitted; thus

$$pH = \tfrac{1}{2}(pK + pC) + 0.217(K/C)^{1/2} \tag{92}$$

If preferable, the correction term can also be expressed in terms of pC and pK as follows:

$$pH = \tfrac{1}{2}(pK + pC) + 0.217 \exp 1.15(pC - pK) \tag{93}$$

Differentiation with respect to pC gives

$$d(pH)/d(pC) = \tfrac{1}{2} + 0.250 \exp 1.15 (pC - pK) \tag{94}$$

where the last term is a small correction which can often be neglected. Numerical values are given in Table IV. The main term in (94) is also obtainable from Eq. (60) in view of Eq. (82).

TABLE IV

Numerical Values of the Correction
Term $\Delta pH = 0.217 \exp 1.15(pC - pK)$
in Eq. (93)

$pK - pC$	ΔpH	$pK - pC$	ΔpH
0.0	0.217	1.6	0.035
0.1	0.193	1.8	0.027
0.2	0.173	2.0	0.022
0.3	0.145	2.2	0.017
0.4	0.137	2.4	0.014
0.5	0.122	2.6	0.011
0.6	0.109	2.8	0.009
0.7	0.097	3.0	0.007
0.8	0.087	3.2	0.005
0.9	0.077	3.4	0.004
1.0	0.069	3.6	0.003
1.2	0.055	3.8	0.003
1.4	0.043	4.0	0.002

The pH course as a function of pC for three weak acids is shown in Fig. 5. The conclusion may be drawn that electrolysis of one single weak acid or base is capable of giving a useful pH gradient comprising about half a pH unit. More extended pH gradients may be obtained by using mixtures of weak acids, as has been shown by Pettersson (1969) and Stenman and Gräsbeck (1972).

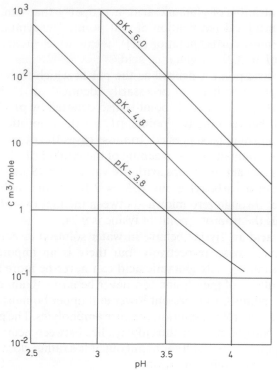

Fig. 5.   pH as a function of concentration for three weak acids, with $pK = 3.8$ (formic acid), 4.8 (acetic acid), and 6.0 (trichlorophenol; cacodylic acid). In an electrolyzer giving rise to linear concentration courses of such acids, nine-tenths of the apparatus will be occupied by a concentration interval of one power of 10. Thus, the pH gradient available within nine-tenths of the apparatus will comprise about 0.5 pH unit. In the remaining one-tenth of the apparatus the pH course will be too steep to be useful.

V. CONVECTION-FREE ELECTROLYSIS OF AN ACIDIC OR BASIC AMPHOLYTE WITH ONLY TWO PROTOLYTIC GROUPS IN THE ISOELECTRIC REGION

A. General

As typical examples of this kind of ampholyte it is suitable to consider glutamic acid and lysine. Glutamic acid has two carboxyl groups, with pK values at 2.19 and 4.25. Owing to the presence of an amino group ($pK = 9.67$) in the molecule, its mean valence passes through zero midway between the first two pK values, i.e., at pH 3.22. This is also expressed by saying that the isoionic point of glutamic acid is 3.22. Rilbe (1973) has suggested the name isoprotic point instead because this pH is the only one that is independent of the concentration of glutamic acid, a property that can be utilized for direct experimental determination of that point. The symbol pI, an abbreviation of the more correct notation $(pH)_i$, is extensively used for the isoprotic point; thus, $pI = 3.22$ for glutamic acid.

The isoelectric point is defined as the pH at which the electric mobility passes through zero; it is not necessarily identical to the isoprotic point. Quite illogically, the isoelectric point is also denoted by pI. When necessary to distinguish between the two points, pI_p for the isoprotic and pI_e for the isoelectric point may be suggested. In most cases the experimental method used leaves no doubt about the meaning of the symbol pI.

Lysine has two amino groups, with pK values at 8.95 and 10.53. Owing to the presence of a carboxyl group ($pK = 2.18$) in the molecule, its mean valence passes through zero midway between the first two pK values, i.e., at pH 9.74. Thus the isoprotic point of lysine is 9.74.

Glutamic acid and lysine behave in water solution essentially as a weak acid and a weak base, respectively, but there is an important difference. Solutions containing only glutamic acid can never be more acidic than pH 3.22, and solutions of pure lysine can never be more alkaline than pH 9.74. The isoprotic points thus represent lower and upper bounds, respectively, to the pH values available to solutions of pure ampholytes. The pH of a solution of a pure ampholyte in water thus always lies between neutral reaction and its isoprotic point. As we shall soon find out, it lies rather close to the isoprotic point already in very dilute solution.

In a solution of pure glutamic acid sufficiently concentrated for the OH^- ions to be negligible, there are three ion species: hydrogen ions H^+, glutamic cations Glu^+, and glutamic anions Glu^-. Their concentrations, denoted by ( ), satisfy the electroneutrality condition

$$(H^+) + (Glu^+) = (Glu^-)$$

However, there are only two transference numbers, that of the glutamate ion constituent and that of the hydrogen ion constituent. Our present problem concerns the possible constancy of the transference numbers of these two ion constituents.

Let us first assume that the mobilities of Glu^+ and Glu^- are the same except for the sign. The net electric mass transport of glutamic acid is then dictated by the concentration difference $(Glu^-) - (Glu^+)$ multiplied by the mobility of the Glu^- ion. The acid then displays the same behavior concerning transference numbers and conductivity as a monovalent weak acid with a dissociation corresponding to said concentration difference and with an anionic mobility equal to that of the ion Glu^-. The reasons for assuming constant transference numbers for glutamic acid are then equally well founded as those for acetic acid.

The assumption of equal absolute mobilities of Glu$^+$ and Glu$^-$, however, is somewhat dubious. The two ions differ in size by two protons, and there may also be large differences in ionic hydration. If the mobilities u_+ and u_- do not add up to zero, a difference will appear between the isoprotic and isoelectric points. It should be remembered, then, that what can be measured by isoelectric focusing is the isoelectric point.

A difference between isoprotic and isoelectric points for glutamic acid must be regarded as probable, but has not yet been demonstrated experimentally. It should be possible to do this by isoelectric focusing since glutamic acid focuses very well in a pH gradient made up of other protolytes. If such a difference really exists, it must be admitted that the transference numbers of the ampholyte are concentration dependent. This is because pH varies with concentration, and the concentration ratio between cations and anions varies with pH.

The possible concentration dependence of the transference numbers of ampholytes cannot be expected to be simple. Since no data are avilable, the best we can do so far is to neglect this concentration dependence. Even an ampholyte will then assume a linear concentration course according to Eqs. (13) and (17) on convection-free electrolysis. The pH gradient obtainable by such experiments will now be computed.

B. The pH of a Solution of a Pure Ampholyte

In this general treatment the ampholyte is denoted by HA, its cationic form by H_2A^+, and its anionic form by A^-. The mass action law for the two dissociation steps on either side of the isoprotic point is

$$hC_0 = fC_+K_1 \tag{95}$$

$$hfC_- = K_2C_0 \tag{96}$$

where h, as before, is the hydrogen ion activity, C_+, C_-, and C_0 are the concentrations of cationic, anionic, and undissociated (or zwitterionic, or both) forms, respectively, K_1 and K_2 are the two dissociation constants, and f is the activity factor for a monovalent ion. If these two equations are solved for C_+ and C_-, respectively, and the solutions inserted into the expression for the total concentration

$$C = C_+ + C_0 + C_- \tag{97}$$

we easily arrive at the following set of equations for the concentrations of the three subspecies:

$$C_+ = h^2C/N^2 \tag{98}$$

$$C_0 = fhK_1C/N^2 \tag{99}$$

$$C_- = K_1K_2C/N^2 \tag{100}$$

where N is a concentration defined by the equation

$$N^2 = h^2 + fhK_1 + K_1K_2 \tag{101}$$

At the hydrogen ion activity where the concentrations of cations and anions are equal, the mean valence of the ampholyte is zero, and consequently we have for the isoprotic point

$$h_i = (K_1K_2)^{1/2}, \qquad pI = \tfrac{1}{2}(pK_1 + pK_2) \tag{102}$$

We further have for the ion product of water,

$$hf(OH^-) = K_w{}^2$$

[Eq. (41)] and to account for electroneutrality,

$$h/f + C_+ = (OH^-) + C_- \tag{103}$$

Elimination of C_+, C_-, and (OH^-) by Eqs. (98), (100), and (41) gives

$$C = \frac{(K_w{}^2 - h^2)(h^2 + fhK_1 + K_1K_2)}{fh(h^2 - K_1K_2)} \tag{104}$$

This equation contains full information of how h, and thus pH, depends on the concentration C of the ampholyte, but it is difficult to interpret since it is of the fourth degree in h. It is not possible by simple means to calculate the pH given by a certain concentration C, but it is easy to calculate the concentration C required to give a certain pH.

It is interesting to note that Eq. (104) becomes easily solvable with respect to h if the total concentration C is eliminated in favor of the concentration C_0 of the uncharged ampholyte by means of Eq. (99). Equation (104) can then be given in the form

$$h^2 = h_i{}^2 \frac{C_0 + K_1(K_w/h_i)^2}{C_0 + K_1} \tag{105}$$

The concentration C_0 increases monotonically with the total concentration C, although in a complicated manner. Equation (105) thus gives the following information:

1.   At very low concentration, the K_1 terms dominate over the C_0 terms, and h approximates K_w, pH is close to 7.
2.   For acidic ampholytes, $h_i > K_w$, hence the numerator is smaller than the denominator, $h < h_i$, pH $>$ pI.
3.   For basic ampholytes, $h_i < K_w$, hence the numerator is bigger than the denominator, $h > h_i$, pH $<$ pI.
4.   It is impossible for h to be equal to h_i, or for pH to equal pI, but pH approaches pI as the concentration increases.

Let us now return to Eq. (104) and make the following substitutions in order to make the equation more convenient to handle:

$$H = h/K_w, \qquad \log H = 7 - \text{pH} \tag{106}$$

$$R = h/h_i, \qquad \log R = \text{p}I - \text{pH} \tag{107}$$

$$M = C/K_w, \qquad \log M = 7 - \text{p}C \tag{108}$$

$$s = f(K_1/4K_2)^{1/2}, \qquad \log s = \tfrac{1}{2}\Delta\text{p}K + \log(f/2) \tag{109}$$

The equation then takes the form

$$fM = \frac{(H^2 - 1)(R^2 + 2sR + 1)}{H(1 - R^2)} \tag{110}$$

In this equation all variables and parameters are dimensionless, which allows the most convenient interpretation. The parameter H describes the deviation from neutrality, R is the deviation from the isoprotic point, M is the concentration with the unit 10^{-7} mole/liter, and s describes quantitatively the distance between the two $\text{p}K$ values on either side of the isoprotic point.

C.   Interpretation

The dependence of fM on pH is given by the derivative

$$\frac{d(fM)}{d(\text{pH})} = \frac{\partial(fM)}{\partial H}\frac{dH}{d(\text{pH})} + \frac{\partial(fM)}{\partial R}\frac{dR}{d(\text{pH})} \tag{111}$$

In this equation we insert the following derivatives, easily derived from Eqs. (106)–(110):

$$\frac{\partial(fM)}{\partial H} = \frac{(H^2 + 1)(R^2 + 2sR + 1)}{H^2(1 - R^2)} \tag{112}$$

$$\frac{\partial(fM)}{\partial R} = \frac{(H^2 - 1)2s(R^2 + 2R/s + 1)}{H(1 - R^2)^2} \tag{113}$$

$$\frac{dH}{d(\text{pH})} = -H \ln 10, \qquad \frac{dR}{d(\text{pH})} = -R \ln 10 \tag{114}$$

Finally, we multiply Eq. (111) by the derivative

$$\frac{d(\text{p}C)}{d(fM)} = -\frac{\log e}{fM} \tag{115}$$

and insert fM from Eq. (110). We thus obtain the equation

$$\frac{d(\text{p}C)}{d(\text{pH})} = \frac{H^2 + 1}{H^2 - 1} + \frac{2sR(R^2 + 2R/s + 1)}{(1 - R^2)(R^2 + 2sR + 1)} \tag{116}$$

The last term in this equation can be split into two fractions as follows:

$$\frac{d(pC)}{d(pH)} = \frac{H^2 + 1}{H^2 - 1} + \frac{1 + R^2}{1 - R^2} - \frac{1 - R^2}{R^2 + 2sR + 1} \qquad (117)$$

This equation, which is surprisingly simple and attractive in view of the complexity encountered on direct differentiation of Eq. (110) after the substitution $R = HK_w/h_i$, shows that log C goes from minus to plus infinity as pH goes from neutrality to pI. This is also illustrated in Figs. 6 and 7.

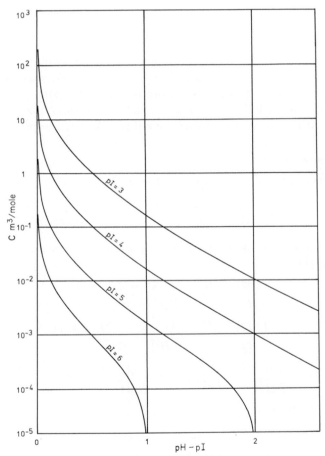

Fig. 6. pH $-$ pI as a function of concentration for ampholytes with an s value of $(10)^{1/2}$ and with isoelectric points at pH 6, 5, 4, and 3. Glutamic acid has pI = 3.22 and s = 5.4 and thus comes fairly close to the uppermost curve. If stationary electrolysis gives rise to a linear concentration course of an ampholyte, which may happen approximately, nine-tenths of the apparatus will be occupied by a concentration interval of one power of 10. The corresponding pH interval will be very small close to pI at high concentrations and less than 1 unit at low concentrations.

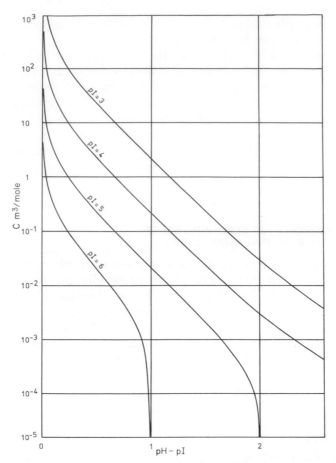

Fig. 7.   pH − pI as a function of concentration for ampholytes with an s value of 100 and with isoelectric points at pH 6, 5, 4, and 3. Arginine has pI = 10.76, corresponding to 3.24 for an acidic ampholyte, and it has s = 52. Its curve thus runs fairly close to the uppermost curve. If stationary electrolysis gives rise to a linear concentration course of an ampholyte, which may happen approximately, nine-tenths of the apparatus will be occupied by a concentration interval of one power of 10. The corresponding pH interval will be very small close to pI at high concentrations and about 0.5 unit at low concentrations.

Although these figures largely speak for themselves, a few comments will be made. For ampholytes with large s values, the last term in Eq. (117) is negligibly small and the equation becomes symmetric with respect to H and R, i.e., with respect to neutrality and pI. Consequently there will be an inflexion at pH = (pI + 7)/2 with a minimum slope of

$$d(pC)/d(pH) = 2 \tag{118}$$

which implies a pH change of half a unit for a 10-fold change in ampholyte concentration. This can be seen in Fig. 7, drawn for ampholytes with $s = 100$.

For ampholytes with small s values, the last term in Eq. (117) is important and approaches the value -1 as s approaches unity. Such ampholytes thus have a minimum slope $d(pC)/d(pH)$ not far from unity, i.e., they exhibit a pH change of about one unit for a 10-fold change in concentration. This is illustrated in Fig. 6, drawn for ampholytes with $s = \sqrt{10}$.

The conclusion can be drawn that single ampholytes can be used for creation of suitable pH gradients within pH intervals comprising up to 1 unit. However, the ampholyte concentrations at which this occurs are very low and the buffer capacity is thus poor.

VI. CONVECTION-FREE ELECTROLYSIS OF A MIXTURE OF TWO AMPHOLYTES

When two ampholytes are mixed in solution, their transference numbers are definitely not constant, and the simultaneous solution of two differential equations such as (9), where both u and E depend on the other ampholyte, is too complicated to be tried. However, with the complete understanding of the behavior of single ampholytes in an electrolysis cell as given by the preceding sections, it is now possible to predict what will happen on electrolysis of a mixture of two ampholytes.

Let us first consider a mixture of one acidic and one basic ampholyte, e.g., glutamic acid and lysine. On convection-free electrolysis these two amino acids will separate completely, so that we will get pure glutamic acid in the anodic and pure lysine in the cathodic part of the electrolyzer. Consequently, the results of Section V apply. At the cathode $pI = 9.74$ of lysine will develop, and pH declines only very slowly toward the middle of the apparatus; at the anode, $pI = 3.22$ of glutamic acid will develop, and pH rises only very slowly toward the middle of the apparatus. At the point where both ampholytes disappear, pH rises very steeply from a low to a high value. Where pH passes through 7, we have pure water with a conductivity so low that it tends to quench the current.

Let us then consider electrolysis of two acidic or two basic ampholytes, e.g., lysine and histidine, isoelectric at pH 9.74 and 7.6, respectively. Lysine as the stronger base will then migrate to the cathode, where a pH slightly less than 9.74 will develop. At this pH, histidine is strongly negatively charged and is therefore repelled by the cathode. It tends to develop a pH plateau at pH 7.6, where it will be essentially pure since lysine is strongly positively charged at that pH and migrates away toward the cathode. Between the two pH plateaus there will be a rather steep pH course and a strong pH gradient. Since there are no acidic ampholytes, the two bases will fill almost the whole electrolyzer, but at the anode pure water will develop with a conductivity so low that it tends to quench the current.

Pure water will not be formed anywhere between the two bases; that is, the separation between them cannot be complete. If this were possible, pH would drop from the vicinity of 9.74 to 7 and then go back again to 7.6. Since negative pH gradients have never been observed experimentally, Rilbe (1973) found reason to formulate the law of pH monotonicity: Convection-free electrolysis always gives rise to a pH course monotonically increasing from anode to cathode. Although this law makes it impossible to separate ampholytes from each other completely, it must be regarded as extremely valuable since negative pH gradients destroy isoelectric separation and since it gives the possibility of getting smooth and useful pH gradients in regions between pI values of individual ampholytes.

VII. CONVECTION-FREE ELECTROLYSIS OF A MIXTURE OF MANY AMPHOLYTES

A. General

It has now been established that every convection-free electrolysis gives rise to a positive pH gradient and that every ampholyte present tends to give rise to a pH plateau where pH changes very slowly, whereas pH changes rapidly in regions where no ampholytes are isoelectric. The question of how to obtain an evenly changing pH course is then understood to be equivalent to the problem of finding a sufficient number of ampholytes with isoelectric points distributed evenly over the pH scale. The pH gradients created in this way have been called natural pH gradients (Svensson, 1961) because they are formed by the action of the electric current, whereas pH gradients made by mixing buffer solutions of different pH values were called artificial pH gradients. The natural gradients have the great advantage of being stable for a very long time (yet not in gels; see Chrambach et al., 1973) on passage of current, whereas the current destroys artificial gradients by its electrolytic separation of acids and bases. However, with big reservoirs of buffer solutions even artificial pH gradients can be made quite stable. In density gradient columns, a natural pH gradient exhibits a slow drift toward the cathode, which phenomenon is still unexplained (Fawcett, 1975).

B. Carrier Ampholytes. Ampholine

Up until now the treatment has been limited to analysis of pH courses obtainable by electrolysis of ampholytes, but more than a suitable pH course must be required for a successful isoelectric separation of proteins. The proteins are themselves ampholytes capable of influencing the pH course. In general, this is undesirable, and therefore one must require that the ampholytes have a sufficiently strong buffer capacity, so that the proteins under analysis get only a minor or negligible influence on the pH course. Another property of great importance is the conductivity. In the preceding sections it has been mentioned several times that electrolyte-free water develops at some point in the electrolyzer on convection-free electrolysis of single substances, that pure water has such a low conductivity that the current is severely quenched, and consequently that the field strength becomes insufficient for effective separation in other parts of the apparatus where there is a much higher conductivity. Thus we must also require that the ampholytes have a sufficient conductivity. This requirement applies especially to those ampholytes which are isoelectric in the neutral range in order to compensate for the deep conductivity minimum at pH 7 exhibited by the water ions. In the steady state all ampholytes are exactly or nearly isoelectric; hence both requirements have to be satisfied in that state.

1.   Buffer Capacity in the Isoprotic State

The mean valence of the ampholyte is given by the equation

$$\bar{z} = (C_+ - C_-)/C \tag{119}$$

and with the aid of Eqs. (98), (100), (101), (107), and (109), we obtain

$$\bar{z} = (R^2 - 1)/(R^2 + 2sR + 1) \tag{120}$$

The isoprotic point is defined as the pH where $\bar{z} = 0$ [cf. Eq. (102)], i.e., by the condition $R = 1$. The molar buffer capacity, being identical to the slope of the titration curve, is the negative derivative of the mean valence with respect to pH. Differentiation of (120) followed by putting $R = 1$ gives

$$d\bar{z}/dR = 1/(1 + s) \tag{121}$$

Further we have, because of Eq. (107),

$$dR/d(\text{pH}) = -R \ln 10 \tag{122}$$

Multiplication and change of sign gives the molar buffer capacity in the isoprotic state

$$B_i = -d\bar{z}/d(\text{pH}) = (\ln 10)/(1 + s) \tag{123}$$

The maximum buffer capacity of a monovalent weak acid or base, i.e., its buffer capacity at pH $= \text{p}K$, is

$$B_1 = (\ln 10)/4 \tag{124}$$

The relative buffer capacity, being defined as the ratio between (123) and

(124), is thus

$$b_i = 4/(1 + s) \tag{125}$$

Since this cannot possibly be more than 2, it follows that s cannot possibly be smaller than unity:

$$s \geq 1 \tag{126}$$

By virtue of Eq. (109), this is equivalent to the known condition for bivalent protolytes:

$$\Delta pK' = pK'_2 - pK'_1 \geq \log 4 \tag{127}$$

where $K'_1 = fK_1$ and $K'_2 = K_2/f$ are the stoichiometric dissociation constants. Equation (125) reveals that the buffer capacity in the isoprotic state decreases with increasing s, i.e., with increasing ΔpK across the isoprotic point.

2.   Conductivity in the Isoprotic State

The conductivity is of course proportional to the degree of ionization α defined by the equation

$$\alpha = (C_+ + C_-)/C \tag{128}$$

By the use of the same equations as before we find

$$\alpha_i = 1/(1 + s) \tag{129}$$

and comparison with (125) gives the relation

$$b_i = 4\alpha_i \tag{130}$$

Buffer capacity and degree of ionization in the isoprotic state thus always occur together. The requirement of good values for these two properties has been found to be equivalent to the requirement of a small pK difference across the isoprotic point. Figure 8 shows how both buffer capacity and degree of ionization vary with the values of ΔpK and s.

Ampholytes with sufficiently good buffer capacity and conductivity in the isoelectric state were called *carrier ampholytes* (Svensson, 1962). Great difficulties were encountered in finding a sufficient number of carrier ampholytes among commercially available chemicals. Not one single carrier ampholyte isoelectric between pH 4 and 7 could be found despite extensive searching. Since a majority of proteins are isoelectric within this range, synthetic work became a necessity. It was very fortunate that Vesterberg (1969) became interested in this problem and had good success with his synthetic procedure, obtaining a very large number of homologs and isomers of polyamino–polycarboxylic acids. These carrier ampholytes are now

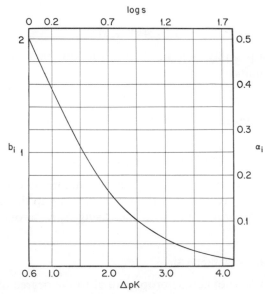

Fig. 8.   Relative buffer capacity b_i and degree of ionization α_i in the isoprotic state as a function of ΔpK and the parameter s [Eq. (109)].

commercially available under the trade name Ampholine through LKB-Produkter AB, Stockholm. The original pH range was 3–10, but it has recently been extended down to about 2.5 and up to about 11.5. Detailed information on the properties of Ampholine is given by Vesterberg in Chapter 3.

C.   Resolving Power

It is now well established that convection-free electrolysis of Ampholine gives very smooth pH courses without any detectable pH plateaus. Thousands of applications have proved this preparation to be extremely useful. We will now discuss what happens to a protein subject to electrolysis together with Ampholine.

In Fig. 9 there are two vertical and two horizontal axes. The vertical axis directed downward represents the coordinate x in an electrolysis column having the anode at the top. The vertical axis directed upward represents the mobility u of a protein undergoing isoelectric focusing. The horizontal axis directed to the right represents pH, whereas the horizontal axis directed to the left represents the concentration C of the same protein in the steady state. The zero point on the x axis has been chosen at the level where the protein is isoelectric and thus has its concentration maximum.

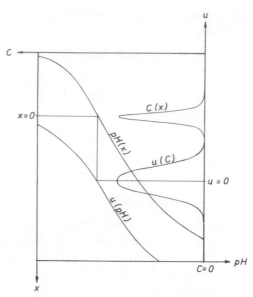

Fig. 9. Illustration of the theory of resolving power. The four axes belong to pH, column coordinate (x), mobility (u), and concentration (C) of the protein. The four curves show pH as a function of x, u as a function of pH, C as a function of x, and u as a function of C.

Between these four variables, four functions can be illustrated with their respective curves, namely pH and C as functions of x, and u as a function of pH and of C. These four functions are shown in the figure. At the level $x = 0$, pH equals pI, and there the function u(pH) passes through zero.

The protein, as all other ampholytes, obeys Eq. (9). According to experimental evidence, protein zones are generally very narrow; i.e., a focused protein zone is confined within a narrow pH interval. Within such an interval the pH gradient d(pH)$/dx$ and the mobility slope du/d(pH) can be regarded as constant. With the x axis pointing in the direction of the current, the pH gradient becomes a positive quantity, but the mobility slope is always negative. Thus p defined as

$$p = -\frac{du}{dx} = \frac{-du}{d(\text{pH})}\frac{d(\text{pH})}{dx} \tag{131}$$

is a positive quantity, and, since $u = 0$ for $x = 0$, the relation between u and x becomes

$$u = -px \tag{132}$$

Inserted into Eq. (9), this gives

$$dC/C = -(pE/D)x\,dx \tag{133}$$

If even E and D can be regarded as constant within the zone, this differential equation has the solution

$$C = C(0) \exp(-pEx^2/2D) \tag{134}$$

where $C(0)$ is the integration constant, which physically means the concentration maximum at the level $x = 0$. If the mass content m of the zone is preferred, the equation becomes

$$C = (m/q)(pE/2\pi D)^{1/2} \exp(-pEx^2/2D) \tag{135}$$

where q is the cross-sectional area. Equations (134) and (135) express a Gaussian concentration distribution with inflection points situated at

$$x_i = \pm(D/pE)^{1/2} \tag{136}$$

Narrow zones are thus favored by a low diffusion coefficient, a high field strength, a large mobility slope, and a strong pH gradient.

Very narrow protein zones are impressive to look at since they give the impression of a good resolving power, but this is not necessarily true. In order to judge the resolving power correctly, it is necessary to consider two neighboring protein zones and to analyze what pI difference between the proteins is necessary for the corresponding zones to be clearly visible as two concentration peaks. A strict definition of a just resolved double zone is then first required. Svensson (1966) has shown that the sum of two identical Gaussian curves situated $3x_i$ apart has a concentration minimum between the maxima as deep as the inflection points (0.61 of the maxima), and he suggested this condition to be adopted as a suitable criterion on a just resolved double zone. The pH difference between two such peaks becomes

$$\Delta(\text{pH}) = 3x_i \, d(\text{pH})/dx \tag{137}$$

Because this is the smallest difference in isoelectric point that can be detected with certainty, the resolving power of the method is obtained by changing $\Delta(\text{pH})$ to $\Delta(\text{p}I)$ and by insertion of x_i from Eq. (136). With the aid of (131) we thus obtain

$$\Delta(\text{p}I) = 3\left(\frac{D(d\text{pH}/dx)}{E(-du/d\text{pH})}\right)^{1/2} \tag{138}$$

It is thus found that a weak pH gradient favors a high resolution, whereas a strong gradient favors narrow zones. Broad protein zones are thus very compatible with a high resolution.

With numerical data pertaining to myoglobin Vesterberg and Svensson (1966) found a theoretical resolving power of 0.02 pH unit, which was in excellent agreement with their experimental results. They found a pI shift of 0.5 pH unit on conversion from ferric to ferrous myoglobin and vice versa. The resolving power thus corresponds to a charge difference per molecule of 0.04 electron charge. A still better resolving power has been demonstrated in connection with isoelectric focusing in gels, probably of the order of 0.01 pH unit or less. This extremely good resolving power explains the very rapid rise in the number of applications of the method.

D.   Analysis and Separation of Peptides

Analysis of peptides is readily made by isoelectric focusing if their mobility slopes are big enough and if their light absorption allows easy detection in the mixture with Ampholine. Purification is more difficult because Ampholine contains material with a molecular weight as large as 5000 g/mole. The separation of a focused peptide from Ampholine may therefore be quite complicated.

Gasparic and Rosengren (1974) have recently reported that a special small-molecular-size Ampholine preparation has been made available. In order to facilitate the study of its properties, it has been labeled with ^{14}C. If this preparation is used as carrier ampholytes, it can subsequently be removed from focused peptides bigger than 2000 g/mole by Sephadex filtration.

As has been pointed out by Catsimpoolas and Campbell (1972), arbitrarily small peptides and even amino acids may be analyzed by isoelectric focusing provided the mobility slope is big enough and detection by some means is possible in the mixture with Ampholine. For purification, however, Ampholine cannot be allowed. Purification of oligopeptides and amino acids without admixture of Ampholine is impossible because of the law of pH monotonicity, i.e., by the considerable overlapping of every two adjacent concentration profiles. Two ways of circumventing this obstacle may be hinted at here.

One possibility consists of allowing a system of proteins available in large amounts, e.g., egg white or bovine serum, to take over the role of carrier ampholytes and to perform analysis of oligopeptides in pH gradients given by the isoelectric condensation of these proteins. For good success, isoelectrically insoluble proteins should first be removed or solubilized. Purification of peptides from such a medium is possible by dialysis of isolated fractions.

The other possibility is restricted to pH values remote from the neutral range and is based on using pH gradients obtainable by electrolysis of weak acids or bases, as has been discussed in Section IV. Complete separation is possible because the weak acid or base will intervene between the peptide zones. If volatile acids or bases are chosen, it will be easy to remove them after separation.

Oligopeptides contain just a few protolytic groups, and it is quite possible that only two of them are active in the isoelectric state; the theory given in Section V is then fully valid. If there are three or four such groups dissociating in the vicinity of pI, the theory loses quantitative validity, but remains true concerning all practical aspects.

With only two protogenic groups active near pI it is meaningful to define the mobility U of the monovalent cation of the peptide, and if the mobility of the monovalent anion is assumed to be $-U$, then the variable mobility in the isoelectric region can be written

$$u = U\bar{z} \tag{139}$$

and insertion of the mean valence from Eq. (120) gives

$$u = U(R^2 - 1)/(R^2 + 2sR + 1) \tag{140}$$

Differentiation with respect to pH and insertion of (122) gives for the isoprotic state, where $R = 1$,

$$du/d(\text{pH}) = -(U \ln 10)/(1 + s) \tag{141}$$

The resolving power for peptides may therefore be written

$$\Delta(\text{p}I) = 3 \left(\frac{D(d\text{pH}/dx)(1 + s)}{EU \ln 10} \right)^{1/2} \tag{142}$$

The ratio D/U does not vary much from one peptide to another. It is seldom known, but can be estimated from available data for compounds of similar size and shape.

E.   Condition for a Stepless pH Course

It has been mentioned previously that each ampholyte tends to develop a pH plateau and that the question of how to obtain a smooth pH course is equivalent to the problem of finding a sufficient number of ampholytes with pI values that are evenly distributed over the pH scale. The question of how many individual ampholytes are contained in the Ampholine preparation per pH unit, and how many are really necessary in order to get an acceptable pH course, has been raised several times. The most serious attempt to answer this question has been presented by Almgren (1971) in his theoretical treatment of systems of ampholytes. The problem can also be attacked in a very simple way by formulating a condition for completely unresolved double zones of adjacent carrier ampholyte peaks. From Fig. 3a of Svensson's (1966) paper, it is evident that the sum of two similar Gaussian curves has one flat maximum and no minimum at all for a peak-to-peak distance equal to or less than two standard deviations. If all individual carrier ampholytes lie as closely spaced as that, all of them are completely unresolved from their neighbors. No plateaus in the pH course can then be expected. The greatest allowable pI difference between adjacent individual carrier ampholytes is thus found by using the factor 2 instead of 3 in Eq. (138) and by insertion of the expression (141) for the mobility slope:

$$\Delta(\mathrm{p}I) < 2 \left(\frac{D(d\mathrm{pH}/dx)(1 + s)}{EU \ln 10} \right)^{1/2} \tag{143}$$

This inequality is in complete agreement with Almgren's conclusions. The dependence on the parameter s is of great interest. This parameter is the quality mark of the ampholytes: The lower the s value, the better are the conductivity and the buffer capacity. Condition (143) now shows that ampholytes of high quality must lie more closely spaced on the pH axis than ampholytes of a lower quality. Ampholytes with low s values are most essential in the pH region round the neutral point in order to compensate for the extremely low conductivity of the water ions in that region.

Almgren's results put in relation to other requirements seem to indicate that high-quality as well as low-quality carrier ampholytes, with a preponderance for the high quality in the neutral range, are essential for successful isoelectric focusing. A mixture of natural amino acids in the form of a protein hydrolysate does not satisfy these requirements at all, because in that mixture the neutral amino acids have very large, the acidic and basic ones very small s values, histidine being an exception. (Histidine is generally accepted as a member of the basic group of amino acids because of its two nitrogen centers, but judging from its isoprotic point of 7.6, it is more neutral than the so-called neutral amino acids, which are isoprotic at pH 6.0.) Nothing quantitative is known regarding the composition of Ampholine with respect to the pI and s distributions, but the numerous successful applications that have been performed with this preparation indicate that it has unexpectedly good properties, at least in view of the great difficulties that have been encountered in its synthesis and purification.

REFERENCES

Almgren, M. (1971). *Chem. Scripta* **1**, 69–75.
Catsimpoolas, N., and Campbell, B. E. (1972). *Anal. Biochem.* **46**, 674–676.
Chrambach, A., Doerr, P., Finlayson, G. R., Miles, L. E. M., Sherins, R., and Rodbard, D. (1973). *Ann. N.Y. Acad. Sci.* **209**, 44–60.
Conway, B. E. (1952). "Electrochemical Data." Elsevier, Amsterdam.
Fawcett, J. S. (1975). *In* "Progress in Isoelectric Focusing and Isotachophoresis" (P. G. Righetti, ed.) pp. 25–37. Elsevier, Amsterdam.
Gasparic, V., and Rosengren, Å. (1974). *Sci. Tools* **21**, 1–2.
Güntelberg, E. (1926). *Z. Phys. Chem.* **123**, 243.
Kunkel, H. G., and Tiselius, A. (1951). *J. Gen. Physiol.* **35**, 89.

Pettersson, E. (1969). *Acta Chem. Scand.* **23**, 2631–2635.

Rilbe, H. (1973). *Ann. N.Y. Acad. Sci.* **209**, 11–22.

Robinson, R. A., and Stokes, R. H. (1959). "Electrolyte Solutions," 2nd ed. Butterworths, London.

Stenman, U. H., and Gräsbeck, R. (1972). *Biochim. Biophys. Acta* **286**, 243–251.

Svensson, H. (1956). *Sci. Tools* **3**, 30–35.

Svensson, H. (1961). *Acta Chem. Scand.* **15**, 325–341.

Svensson, H. (1962). *Acta Chem. Scand.* **16**, 456–466.

Svensson, H. (1966). *J. Chromatogr.* **25**, 266–273.

Tiselius, A. (1941). *Sv. Kem. Tidskr.* **53**, 305–310.

Vesterberg, O. (1969). *Acta Chem. Scand.* **23**, 2653–2666.

Vesterberg, O., and Svensson, H. (1966). *Acta Chem. Scand.* **20**, 820–834.

3 THE CARRIER AMPHOLYTES

Olof Vesterberg

Chemical Division
Department of Occupational Health
National Board of Occupational Safety and Health
Stockholm, Sweden

I. DESIRABLE PROPERTIES

As has been mentioned, the principle of isoelectric separation was tried for the fractionation of amino acids and proteins, although with very limited success. Practical difficulties cropped up in the forms of very uneven field strength and unsuitable pH distribution between the electrodes. Kolin (1954) evolved an approach to solve this. However, other problems then appeared, e.g., instability of pH gradients with time. The methodology was very significantly advanced when Svensson (1961, 1962), through theoretical calculations, presented a hypothesis for isoelectric focusing of amphoteric molecules.

In order to explain the principle of the method let us first start with an electrolysis cell containing a water solution of sodium sulfate. The sodium ions will be transported toward the cathode where they will yield NaOH and a high pH. The sulfate ions will be transported toward the anode where they will yield H_2SO_4 and a low pH. Pure water will be obtained in between with a very low conductivity κ. The current i will be influenced by the cross-sectional area q of the cell. Since the field strength $E = i/\kappa$ (Svensson, 1961) a very large part of the voltage will be present in the middle of the cell and very little will be left in other parts. Let us now add an amino acid to the system. At neutral pH the amino acid will exist in its zwitterionic form, the amino group of which is protonated and carries a positive charge, and the carboxylic group of which is dissociated and carries a negative charge. These groups have dissociation constants pK_2 and pK_1, respectively:

$$\underset{\substack{(C_+) \\ \text{in acidic pH}}}{NH_3^+\!-\!CH_2\!-\!COOH} \;\underset{+H^+}{\rightleftarrows}\; \underset{(C\pm)}{NH_3^+\!-\!CH_2\!-\!COO^-} \;\overset{-H^+}{\rightleftarrows}\; \underset{\substack{(C_-) \\ \text{in alkaline pH}}}{NH_2\!-\!CH_2\!-\!COO^-}\;(2)$$

In the electric field the molecules will be transported to a pH where the net charge is zero, i.e., the isoelectric point (pI) where pH = pI. Between pH 3.5 and 10.5 the water ions can be disregarded.

If C_+ is the concentration of the cationic form, C_- the concentration of the anionic form, and C the total concentration, then the mean valence is given by the equation

$$\bar{Z} = (C_+ - C_-)/C \tag{1}$$

and the degree of ionization by

$$\alpha = (C_+ + C_-)/C \tag{2}$$

If the two dissociation constants are K_1 and K_2, then the isoelectric point is given by the equation

$$pI = \tfrac{1}{2}(pK_1 + pK_2) \tag{3}$$

Svensson (1962) and Rilbe (1973a) have shown that the degree of ionization in the isoelectric state is

$$\alpha_i = 2/(2 + 10^{pI - pK_1}) \tag{4}$$

and that the buffer capacity

$$B = -d\bar{Z}/d(\text{pH}) \tag{5}$$

in the isoelectric state is given by

$$B_i = (2 \ln 10)/(2 + 10^{pI - pK_1}) \tag{6}$$

Because the conductivity rises with the degree of ionization the conductivity contribution of an ampholyte at pI is proportional to α_i. From this it can be understood that there exists a positive correlation between the buffer capacity and the conductivity.

Division by the maximum buffer capacity of a monovalent weak acid, $(\ln 10)/4$, gives the relative buffer capacity in the isoelectric state:

$$b_i = 8/(2 + 10^{pI - pK_1}) = 4\alpha_i \tag{7}$$

A good buffer capacity is thus always connected with a good conductivity, and both are good if $\Delta pK = pK_2 - pK_1$ is relatively small.

From these equations it can be seen that the smaller the difference $\Delta pK = pK_2 - pK_1$, and consequently also $pI - pK_1$, the larger is the buffer capacity. In Fig. 1 the relationship of ΔpK with the buffer capacity in units of the maximum capacity of a monovalent weak protolyte, i.e., the relative buffer capacity b_i, can be seen by using the coordinate in this figure. Here it can be seen that large numerical values are obtained for small difference between pK_2 and pK_1, and that as this difference increases the buffer capacity diminishes. When $pK_2 - pK_1 = 4$ there remains only about 4% of the maximal buffer capacity. At $pK_2 - pK_1 = 2.5$ the ampholyte has about 20% of its maximal buffer capacity, which is quite good. From this it can be understood that in order to obtain a certain minimal buffer capacity the ΔpK distance should preferably not exceed a certain value. It is well known that the buffer capacity of a protolyte and accordingly an ampholyte can be calculated from the slope of a titration curve (cf. Fig. 2). Here it can be seen that glycine with a ΔpK of 7.4 is unsuitable as

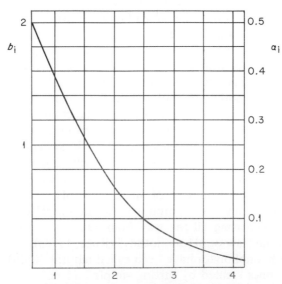

Fig. 1.   Relative buffer capacity b_i in units of the maximum capacity and degree of ionization a_i at pI. [From Rilbe (1973a), used with permission of the New York Academy of Sciences.]

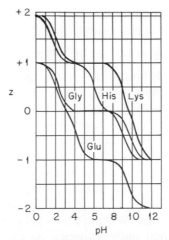

Fig. 2.   Titration curves of lysine, histidine, glycine and glutamic acid. A high slope is equal to a good buffer capacity, i.e., as seen close to the pK values, whereas a low slope or a more horizontal part indicates a low buffer capacity, i.e., as seen at some distance from the pK values (cf. Fig. 1). The pI is situated between two pK values. It can be seen that glycine has a very low buffer capacity at pI of about 6 because the pI − pK difference is large. On the other hand, lysine, histidine, and glutamic acid have much higher buffer capacities at pI values of 9.7, 7.5, and 3.2, respectively. [From Svensson (1962).]

a carrier ampholyte since it has an extended horizontal part of the titration curve about 2 pH units on either side of its pI at pH 6.

The theory and examples referred to here are only valid for simple ampholytes with two pK values in the vicinity of the pI. For ampholytes with more pK values near pI more than two protolytic groups may be active at pI, which increases the conductivity and buffer capacity.

If we introduce another ampholyte into the system, it will also be transported in the electric field to its pI, where it will dictate pH if it has a certain buffer capacity and is present in sufficient amount. If other ampholytes with different pI values are also introduced into the system, they will arrange themselves in a series: The most acidic one will be closest to the anode; the higher the pI value, the closer will the ampholyte be to the cathode. Therefore, with a small number of ampholytes pH steps can be noticed. With an increasing number of ampholytes a smooth pH gradient is obtainable, with pH increasing all the way from anode to cathode. This is the implication of the "law of pH monotony" (Rilbe, 1973a). The pH gradients obtainable with ampholytes have been called natural, to differentiate them from artificial ones created by mixing simple buffers.

Svensson (1961) presented an equation describing the concentration of an electrolyte as an equilibrium between electric mass transport and diffusional mass flow:

$$Cui/q\kappa = D(dC/dx) \tag{8}$$

where C is the concentration of the component, u is the electric mobility $(cm^2\ V^{-1}\ sec^{-1})$ of the ionic constituent, i is the electric current (A), q is the cross-sectional area (cm^2) of the electrolytic medium, measured perpendicularly to the direction of the current, κ is the conductivity at the point of focusing $(ohm^{-1}\ cm^{-1})$, D is the diffusion coefficient (cm^2/sec) of the component, and x is the coordinate along the direction of the current, and thus increases from 0 at the anode toward the cathode (cm).

It was also shown by Svensson (1961) that there exists a relationship between the field strength and the concentration distribution of the ampholyte which can be regarded as a bell-shaped Gaussian curve with the inflection points at

$$x_1 = \pm(D/pE)^{1/2} \tag{9}$$

where E is the field strength and

$$p = -\frac{du}{d(\text{pH})}\frac{d(\text{pH})}{dx} \tag{10}$$

From this equation some very important conclusions can be drawn:

1. The degree of focusing is directly proportional to the square root of the field strength E at the point of focusing. Furthermore, since $E = i/q\kappa$, the conductivity course is as essential as the pH course. For general applicability of the method and in order to obtain a fairly constant field strength in the main part of the electrofocusing cell, we must have many ampholytes with different pI values in the range between pH 2 and 11, and which possess a certain conductivity in the isoelectric state. Ampholytes used for this purpose have been called carrier ampholytes (Svensson, 1961).

2. Proteins are especially suitable for isoelectric focusing, partly because of their steep electrophoretic mobility curves [large values of $du/d(\text{pH})$] at pI, and partly because of their low diffusion coefficients.

A consequence of the general theory is that ampholytes will show overlapping bell-shaped curves in the steady state. Two proteins can be completely separated only in the presence of another ampholyte having a certain buffering capacity. It is not necessary that the ampholyte has a pI between those of the proteins to be separated, since the ampholytes show overlapping bell-shaped concentration distributions (Svensson, 1961; Vesterberg, 1968). The ideal carrier ampholytes must fulfil certain criteria. They must possess a certain buffer capacity in the isoelectric state to be able to determine the pH course also in the presence of proteins. As already mentioned, they should also give a certain conductivity contribution.

Another important property of chemicals to be used as carrier ampholytes is sufficient solubility in water. In practice the average concentration used is often 1 or 2% (w/v).

In order to facilitate detection of proteins after focusing in the carrier ampholyte mixture it is helpful if the ampholytes have a low light absorption at 270 nm or at longer wavelengths to permit the use of this versatile principle. In order to decrease the risk for hydrophobic interaction and binding with proteins the ideal carrier ampholytes should not contain hydrophobic groups. It is well known for some proteins, e.g., albumin, that binding of this type with small molecules can be relatively strong. Most of the ampholytes listed by Svensson (1962) do not fulfill the given requirements. Furthermore, no satisfactory ampholytes with pI values between 3.9 and 7.3 were found. The carrier ampholytes should also be of low molecular weight. This is partly because the carrier ampholytes should not focus too well. As can be seen in Eq. (9) this requirement is partly satisfied for molecules with a relatively high value of D, i.e., molecules small in comparison with proteins. Furthermore, the fractionated proteins are contaminated by the carrier ampholytes, but if the carrier ampholyte molecules are not too big, they may be easily removed by molecular sieving processes, e.g., dialysis or gel filtration on Sephadex.

An extensive search for commercially available chemicals which might be useful as carrier ampholytes was not successful before development of the new system of carrier ampholytes.

II. SYNTHESIS OF A SYSTEM OF CARRIER AMPHOLYTES

Knowing the criteria for good carrier ampholytes, we could consider synthesizing "tailor-made" ampholytes by incorporation of an acidic and a basic group with closely spaced pK values into one molecule. However, this approach is fraught with difficulties. First, there are only very few monovalent protolytes with dissociation constants (pK values) in the pH region 5–9. Second, when incorporating an acidic and a basic group in a molecule, the pK values of the resulting ampholyte generally turn out to be far from the pK values of the original molecules. At first this was regarded as an obstacle, but later on a related phenomenon was taken advantage of (Vesterberg, 1968, 1969b). The interaction between protolytic groups within the same molecule can be studied for some polyethylene polyamines in Table I. The amines containing four or more amino groups have pK values distributed mainly in the interval between 3 and 10. By coupling to these amines one or more residues containing a carboxylic group, many ampholytes with different pI values can be obtained (Vesterberg, 1968, 1969b). For a general formula see Fig. 3. The carrier ampholytes (CAs) are homologs of aliphatic polyamino–polycarboxylic acids. Since the acid residues can be situated at different sites

TABLE I

Dissociation Constants for Ethylene Amines[a]

Amine	pK_1	pK_2	pK_3	pK_4	pK_5	Temp. (°C)
Ethylenediamine	10.1	7.0				20
Diethylenetriamine	9.9	9.1	4.3			20
Triethylenetetramine	9.9	9.2	6.7	3.3		20
Tetraethylenepentamine	9.9	9.1	7.9	4.3	2.7	25

[a] From Vesterberg (1969b).

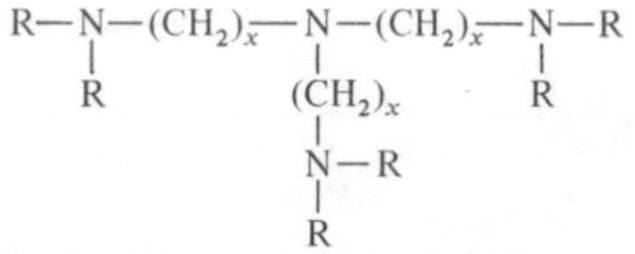

Fig. 3.   A schematic formula illustrating the CAs which are aliphatic polyamino–poly-carboxylic acids. x is generally 2, 3, or 4; R is H or —$(CH_2)_x$N—R(R), or —$(CH_2)_y$—COOH, where y is generally 1 or 2.

in the amine molecule, many isomers are also obtained which have slightly different pK values due to somewhat different mutual interactions. This is very valuable because isomers increase the number of ampholytes obtainable.

Because many CAs with different pI values are needed, it is advantageous to use a synthesis method by which many ampholytes are obtained in one or a few procedures. This means that the CAs are obtained as a mixture. They are actually used for protein separations in a mixture, and sometimes after coarse fractionation. It was of primary interest to separate the synthesis products in order to assess their properties. Fractionation of the ampholytes according to their pI was achieved in a multicompartment electrolyzer, which is an improved type described earlier in the literature (Tiselius, 1941). Further description of the apparatus has been published (Vesterberg, 1969b).

However, it is theoretically impossible to separate completely by iso-electric focusing CAs with closely spaced pI values. The observation from the fractionation experiments was also that the ampholytes are merely separated into groups having pI values within a narrow pH range. Valuable information has been obtained by analysis of such fractions. The theoretically predicted properties of the CAs were found to be valid. The resulting pK values, which are very important for the electrochemical properties of the ampholytes, were found to be predictable to a high degree (Vesterberg, 1968, 1969b). The pI values of the ampholytes first obtained were found to be in the pH range between the pK of the most acidic carboxylic group at about pH 3 and the pK of the most alkaline amino group at about pH 10. For an illustration of titration results in simple cases, see Fig. 4.

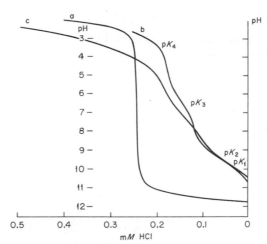

Fig. 4. Acidimetric titration curves with 1.0 M HCl of (a) 0.25 ml of I M NaOH, (b) triethylenetetramine, (c) triethylenetetramine after addition of propionic acid residues and after isolation by electrolysis as described by Vesterberg (1969b) of a fraction with pH 5.5 estimated to contain ampholytes with three acidic residues. The amount of ampholyte used for titration was estimated to be 30.6 mg. The buffer capacity at pH values close to pI ~ 5.5 was calculated to be 0.82 μequiv/mg pH. Note the pK values of the amine indicated in the figure. For the numerical values of the pKs of the amines (b) see Table I. [From Vesterberg (1973a), used with permission of the New York Academy of Sciences.]

In order to extend the pH range other synthesis procedures had to be developed. The polyethylene polyamines mentioned earlier (cf. Table I) do not have pK values high enough to serve as starting materials for synthesis of ampholytes with pI values above pH 10. Knowing the factors influencing the pK values of amines, it was logical to scrutinize aliphatic amines having amino groups more than three methylene groups apart. Some of these amines have pK values high up in the alkaline region (cf. Table II). It was found that by coupling molecules with carboxylic acid groups to such amines, CAs with pI values even above 11 were obtainable. Furthermore, it was found that the pI values measured after focusing experiments agreed very well with the values calculated from the pK values (Vesterberg, 1973a). Although the guanidine group is strongly basic such compounds were abandoned because it was found that they were not stable in alkaline solution.

To extend the pH range on the acidic side, the interest was again focused on the pK values and the mutual influence of carboxylic groups as well as other groups. As an illustration it can be mentioned that when two carboxylic groups are close to each other in a molecule the mutual interaction results in strong effects on the pK values, as can be seen in Table III. The introduction of one or more amino groups into such molecules results in ampholytes.

TABLE II

Dissociation Constants for Aliphatic Amines and Isoelectric Points for the
Carboxylic Acid Derivatives Thereof[a]

Amine	pK_1	pK_2	Temp. (°C)	Calculated pI at 4°C	Measured pI at 4°C
1,3-Diaminopropane	8.6	10.6	20		
	9.0	10.9	10		
	9.2	11.1	4[b]	10.15	10.15
1,4-Diaminobutane	9.4	10.8	20		
	9.7	11.2	10		
	9.9	11.4	4	10.65	10.67
1,5-Diaminopentane	10.0	11.0	20		
	10.5	11.5	4	11.0	10.95
1,6-Diaminohexane	10.0	11.1	20		
	10.4	11.5	10		
	10.6	11.7	4	11.15	11.08

[a] A comparison of values of calculated isoelectric points and those measured by isoelectric focusing. From Vesterberg (1973a), used with permission of the New York Academy of Sciences.
[b] All pK values at 4°C have been obtained by extrapolation.

TABLE III

Dissociation Constants[a] at 25°C for
Some Dicarboxylic Acids with the
General Formula HOOC—$(CH_2)_x$—COOH

Acid	Where x is	pK_1	pK_2
Malonic	1	1.8	5.7
Succinic	2	4.2	5.6
Glutaric	3	4.3	5.4

[a] From "Organic Chemistry" by D. J. Cram and G. S. Hammond. Copyright 1964, McGraw-Hill Book Co. Used with permission of McGraw-Hill Book Co.

However, much synthesis work had to be done before it was possible to get enough molecules in reasonable yields to cover the pH range 2.5–4, retaining the principle that the acidic CAs should also be aliphatic aminocarboxylic acids. Practically successful separations have also documented the value of the acidic ampholytes (Vesterberg, 1973b).

When dealing with proteins with unknown pI values or with proteins having widely different pI values, it is necessary to use CAs with pI values covering a wide pH range (2.5–11). For more detailed studies of proteins at a high degree of resolution, it is of interest to separate the CAs into groups which can dictate a pH course covering essentially only one or a few pH units. This depends on the fact that the separability of proteins and the precision of the pI determinations are correlated with the shallowness of the pH course, as has been shown (Svensson, 1962; Vesterberg and Svensson, 1966). Such fractionation of CAs can be done in the multicompartment electrolyzer and also with other apparatus for isoelectric focusing, e.g., in Sephadex gels according to Radola (1973) and in density gradient columns (Vesterberg, 1971). A lot of work has been devoted to improving the properties and to optimizing the CA synthesis and fractionation procedures. A large part of this work has been conducted successfully by LKB-Produkter AB and Aminkemi AB (Stockholm, Sweden).

The most important properties and advantages of these new CAs are given here.

1. They are obtained as mixtures of many homologs and isomers with different pK and pI values together covering the pH range 2.5–11.
2. Owing to a good buffer capacity and different pI values, they can dictate pH courses in the pI range of most proteins.
3. The conductivity distribution as caused by them, and thus also the field strength between the electrodes during isoelectric focusing, is satisfactory.
4. Their solubility in water is very good.
5. Their light absorption is low at 270 nm or at longer wavelengths (Davies, 1970; see also Fig. 5). This is important if light absorption measurement is to be used for the detection of proteins focused in the pH gradients.

Numerous applications in focusing and separation of proteins made possible with these polyamino–polycarboxylic acids have demonstrated the value of these CAs. (For references see the *Acta Ampholine* yearly reference list, LKB, Bromma, Sweden.)

III. PHYSICOCHEMICAL PROPERTIES: BUFFER CAPACITY AND CONDUCTIVITY

It has previously been stated that the CAs should have a certain conductivity and buffering capacity at pI, i.e., where the ampholyte is found at focusing. When a mixture of many species is present, as is the case for the currently used CAs, it is very difficult to determine the properties for each individual species because the ampholytes must first be separated. This is understandable since titration and conductivity measurements give the net result. In order to check the properties of the CAs many experiments have been conducted on simple mixtures with relatively few molecular types after

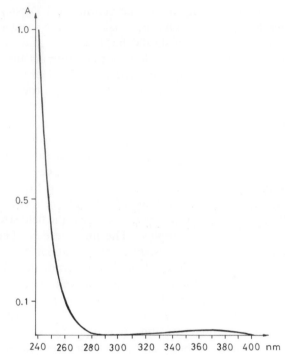

Fig. 5. Light absorption of Ampholine, pH range 3.5–10, in 1% water solution. A is the absorbance in a 1-cm cell. (Courtesy of V. Gasparic.)

fractionation (Vesterberg, 1969b, 1973a). Of the measurements it is easiest to evaluate the buffer capacity at various positions in the pH gradient. Here only the slope of the part of the titration curve that is close to the pH of the fraction obtained after isoelectric focusing is of interest. Buffer capacity at pH values remote from pI and conductivity measurements on crude mixtures are useless. Conductivity measurements are not acceptable because it is difficult to avoid some mixing of adjacent individual CAs at fractionation. This will cause readings that are too high and because of this earlier published values should be interpreted cautiously, especially when complex mixtures are studied. When conductivity has been measured on fractions containing sucrose, as used for the density gradient method, the influence of this must also be considered (Vesterberg, 1973a). If it is desirable to increase the buffer capacity or the conductivity, it is possible to do this by increasing the concentration of the CAs of the corresponding pH region. This is understandable since the mentioned parameters are directly concentration dependent. This is part of the explanation for making a cocktail of Ampholine of different pH regions for isoelectric focusing in polyacrylamide gel (Vesterberg, 1972). We still practice this principle. One may ask what the ideal values of the buffer capacity and conductivity should be. It can be said that the buffer capacity should be high enough even locally to counteract the tendency for closely spaced proteins to overlap (Vesterberg, 1973a). Here the buffer capacities of proteins and CAs may be compared. Although the latter generally have much higher values than proteins on a weight basis it must be remembered that proteins can attain a local concentration in a focused zone of several percent. This is one of the reasons why concentrations of CAs that are too low should be avoided. Furthermore, it can be said that a comparatively low conductivity in a certain pH region is not necessarily a drawback. Here one should observe the field strength and ask if the attainable field strength is sufficient where the proteins of interest are located. Generally, the answer will be yes. If not, it is possible to increase the field strength locally by increasing the total voltage applied. The limit here is often set by the power supply and the cooling efficiency of the apparatus. The density gradient, which is very often utilized in density gradient columns, has a limited capacity to dissipate Joule heat locally as well as to counteract convection that may result when local heat production is too high. However, this is merely a question of design, and Rilbe (1973b) has shown that the problems can be circumvented by using a new type of short density gradients which have a considerable stability and also a very high capacity when the maximum amount of protein is considered. Here can also be mentioned the procedure for focusing in flat beds of polyacrylamide gel with good cooling (Vesterberg, 1973c). In every part of such pH gradients very narrow protein zones are obtainable, which is a good indication that the field strength distribution is satisfactory.

When extreme pH regions are involved, difficulties can occur. Here it must first be remembered that above pH 10.5 and below pH 3.5 the abundance and high mobilities of the water ions give an appreciable conductivity. If we wish to extend beyond this pH range, we may obtain imbalance in the distribution of the field strength in the cell. To counteract this CAs around neutrality are often added (Haglund, 1971; Vesterberg, 1973c). It is important to have a very low salt content in the protein solution at studies in extreme pH regions (Vesterberg, 1971). This is because salt ions migrate to the electrodes and give rise to a wide pH gap, especially between the cathode and the acid pH gradient, or between the anode and the alkaline pH gradient when salts are used. It can be concluded that when working in extreme pH ranges there seems to be a need for optimization of the experimental design, in connection with the Ampholine pH ranges used and the proportion of Ampholine to be added.

IV. NUMBER OF CARRIER AMPHOLYTE
SPECIES WITH DIFFERENT p*I*

One may ask how many species with different p*I* are required? When answering this question many factors have to be considered. First, which slope of the pH gradient, i.e., which value of the derivative $d(\text{pH})/dx$, is of interest? Generally it can be said that the narrower the pH range the more CA species are required per pH unit to obtain a relatively even distribution of the buffering capacity and conductivity along the pH gradient. Let us use these two last-mentioned factors as parameters. The degree of focusing of ampholytes is, according to Eqs. (9) and (10), dependent on the field strength. Thus, the higher the field strength, the more species are required. However, the implication of the square root must be considered, e.g., doubling the field strength causes an increase of a factor of only 1.4. Furthermore, this is also not strictly true because as the focusing of the CA increases, the higher the concentration at the maxima will be, which in turn will result in an increased conductivity, and thus according to Eq. (9) a decrease in the local field strength. Thus, in practice the whole system has a built-in self-regulating mechanism that tries to level out such differences. Furthermore, here the relative amounts of the CA species must be taken into consideration. Because we are discussing different molecules with various physicochemical properties it is irrational to compare them on a weight basis. Comparison on a molar basis might be better. However, because it cannot be assumed that a mole of one CA species has the same buffer capacity, mobility, and diffusion coefficient as another CA species, it is even of limited value to say that a CA mixture with equimolar concentrations of each species would be the ideal. Because of these facts attempts to perform a quantitative evaluation of the relative amounts of the Ampholine species by the "caramelization reaction" (Felgenhauer and Pak, 1973), which must be considered to have a very low quantitative ability, are useless and also misleading.

Theoretical calculations have been carried out concerning how many CA species would be required (Almgren, 1971). However, these merely have a philosophical interest rather than a practical applicability because the calculations were based on many extreme theoretical assumptions, such as all species being present in equal amounts and having equal mobilities. Knowing the starting chemicals and the synthetic reactions that are used it has been calculated that more than 360 homologs and isomers of CA species are obtainable for the pH range 3–11 (Vesterberg, 1973a). Because of the large number of species per pH unit it is in practice very difficult to separate all the species in order to check their individual properties. However, by isotachophoresis, which is one of the best resolving methods for such molecules, it has been shown that there are on the average at least 20 different species per pH unit in the Ampholine mixtures in current use (Evererts, personal communication). Thus there is experimental evidence supporting the actual presence of a large number of individual species in the synthetic mixture. Experiments with radioactively labeled molecules have also indicated that there is really a large number (Vesterberg, 1973a). Estimations with refractometry have also supported this as well as shown that the concentration distribution after focusing is rather even. The latter circumstance can most probably be ascribed to the aforementioned self-regulating mechanism. Thousands and thousands of experiments with many different proteins have verified the usefulness of the present CA system and the users of the technique do not have to be overly concerned about the number of different CA species. Finally, let us add that if the buffer capacity should under some special experimental condition turn out to be too low locally, it can be increased simply by adding more CA, which will make an overall increase in the buffer capacity as was indicated earlier.

V. LIGHT ABSORPTION

As mentioned earlier the detectability of proteins after focusing is facilitated by a low absorbance of the carrier ampholytes, especially at 280 nm (cf. Fig. 5). When considering the large number of individual species present in the CA mixture currently used it is rather astonishing that the absorbance of a 1% solution is below 0.05 at wavelengths longer than 280 nm. However, small absorbance peaks can be detected with different pI values, especially when scanning during focusing. This can be explained by the fact that many of the "impurities" have a tendency to focus in narrow zones, thus giving rise to a certain light absorption although the total content may be quite low. However, it is very important to make light absorption measurements free of artifacts as has been pointed out by Rilbe (1973b) and Jonsson et al. (1968). Differences in refractive index as caused by CA zones should not be mistaken for light absorption of impurities or proteins. Furthermore, it has been noticed that the absorbance close to the electrodes, especially the anode solution, can show fairly high values, which may be due to impurities in the stabilizing media, e.g., sucrose, decomposition, or reaction products formed as "Schiff bases." It must be remembered that sucrose has a limited stability at extreme pH, especially in acid, and furthermore that oxygen and other reactive products are formed at the anode, which may spread by diffusion into the electrolysis cell. For this reason chloride ions should be avoided in the protein sample because this gives rise to chlorine at the anode. To counteract some of these factors some advice could be given:

1. The purity and chemical stability of the stabilizing media are important. Sorbitol seems to be preferable to sucrose.

2. The design of the apparatus and especially the anode compartment as well as the fact that extended time usually increases the unwanted effect caused by reactive products should be considered, and unnecessarily long times avoided. This is also very important to prevent undesirable reactions with proteins (Vesterberg, 1973a).

3. Recording the light absorption in a "blind" experiment without protein sample facilitates comparison and makes possible the evaluation of which light absorption refers to protein (Rilbe, 1973b).

VI. DETERMINATION OF pI OF PROTEINS IN THE PRESENCE OF CARRIER AMPHOLYTES AND OTHER ADDITIVES

One outstanding feature of the isoelectric focusing method is the possibility of determining the pI of proteins in a simple way by simply measuring the pH in the fraction where the protein occurs in maximal concentration. Usually these pH values are determined manually. However, the use of flow through pH cells has been reported by Jonsson *et al.* (1968).

There are reasons to believe that the pI value is close to the isoionic point of proteins (Vesterberg, 1968, 1971; Haglund, 1971; Rilbe, 1973a). Unfortunately, published reports on the determination of the isoionic point by an independent method in order to make a comparison possible are very few. The statement seems to hold for deoxyribonuclease (Vesterberg, 1968) and carbonic anhydrase (Jonsson and Pettersson, 1968) on which determinations have been performed.

The pI value of a protein as ascertained by isoelectric focusing seems to be determinable with high precision (Haglund, 1971). This pI value is therefore as important a characteristic of a protein as its molecular weight. In order to be useful the values should be determined at controlled conditions. Because the dissociation constants of most protolytic groups in proteins show a temperature dependence which can influence the pI, it is important to state the temperature of the pH measurement used (Davies, 1970; Vesterberg, 1971). Most investigators have preferred $+4°C$, and a few $+25°C$. Other aspects of pH measurements, especially after focusing in gel, have been taken up at a round table discussion at an international meeting on isoelectric focusing in Milan in 1974 (Vesterberg, 1975). To exclude artifacts, I recommend that the same temperature be used at focusing as at pH measurement. For the temperature dependence of the pH for the CAs, see Davies (1970).

It seems as if the pI obtained is not influenced by the CA concentration when this is kept at reasonable levels (Vesterberg and Svensson, 1966). However, the lower the ampholyte concentration, the longer is the response time in the pH measurement, and the higher is the risk for unwanted influence of pH by external factors, e.g., absorption of CO_2 and impurities. For this reason, an amount corresponding to an average concentration of at least 1% (w/v) of carrier ampholytes is recommended.

An interesting question is the influence of sucrose or other solutes used to make up the density gradient. There is evidence which seems to indicate that the solute concentration generally does not influence the pI value obtained (Vesterberg and Svensson, 1966; Johnsson et al., 1968). However, some dependence could be expected from theoretical considerations. The interesting question is, how big can the influence of the solute be? If the solute concentration is below 40% (w/v), practical experience has shown that the influence is probably in many cases below 0.1 of a pH unit. However, at higher concentrations and at extreme pH a larger deviation can be expected. It has thus been shown that at very alkaline pH the pI values obtained for the multiple molecular forms of cytochrome c were influenced by the sucrose concentration (Vesterberg, 1968). This observed effect could be caused partly by (a) ionization of sucrose at the high pH or (b) instability of the pH gradient. Later experiments have shown that if the sucrose is replaced by, e.g., sorbitol or glycerol, which are not ionized at the pH in question (pH 10.5–11), the solute effect is much less and, furthermore, a pI is obtained which is very close to the one determined by electrophoresis (Vesterberg, 1968). For the reason mentioned, sucrose is less suitable above pH 10. Here it is of interest to mention experiments conducted in density gradients made up of water, H_2O, and D_2O, where pI values were found to be 0.1 of a pH unit higher than those obtained in sucrose density gradients (Fredriksson and Pettersson, 1974). The net result of the following two effects could be responsible for this: The intrinsic dissociation constants in proteins are higher in D_2O than in H_2O, and the pH meter readings in D_2O are lower than those in H_2O.

Another solute that is used quite often is urea, which is usually added in concentrations up to 6 M to increase the solubility of proteins (Vesterberg, 1970). It has been found that a pI value that is several tenths of a pH unit higher may be obtained in urea solutions (Ui, 1973). The explanation for this may be compared with the influence of D_2O on proteins and CA, which may increase the pI for proteins and CA molecules as well. It is also well known that urea may change the structure of a protein, i.e., by unfolding. Thus, another explanation for a difference in pI may be altered secondary and tertiary structures, which may change the mutual influence of dissociating groups, in turn resulting in a shift in pI. Because the effect of urea can be expected to be different for proteins no general correction factor can be applied. In order to compile more data on the influence of solutes on the pI it would be worthwhile for investigators to estimate the solute concentration at the level of the protein and to publish this information.

Another interesting point is the comparison of the pI values estimated by isoelectric focusing with those obtained by electrophoresis. A difference between the pI values determined by both methods is to be expected because it is well known that the pI determined by electrophoresis is dependent on the kind of buffer ions and the ionic strength used (Vesterberg, 1968). Still, some investigators get confused when the values do not agree. It can be stated that the pI values for a protein may differ up to some pH units if the determinations are made by electrophoresis in different kinds of buffers or at different ionic strengths. Usually the pI values are higher the lower the ionic strength, which has been explained by the binding of anions to proteins at electrophoresis (Tiselius and Svensson, 1940). At isoelectric focusing the ionic strength is low. From this it follows that isoelectric focusing yields pI values that are generally higher than those found with electrophoresis (Vesterberg, 1968, 1973a).

VII.   STABILITY OF PROTEINS AND USE OF ADDITIVES

It seems as if most proteins tolerate the CAs very well. A protective effect on susceptible proteins has even been observed in some cases, as already mentioned. As aminocarboxylic acids they seem to provide a stabilizing environment for proteins similar to that provided by amino acids (Vesterberg, 1968, 1973a). Furthermore, the CAs have a definite solubilizing effect on proteins which can be compared with that of amino acids (Vesterberg, 1968, 1973a). Polyacrylamide gel also seems to have a solubilizing effect on some proteins. Moreover, isoelectric focusing is often carried out in solutes such as sucrose or glycerol which also have a documented stabilizing effect on proteins (Vesterberg, 1968, 1973a; Miner and Heston, 1972). When problems occur with proteins that have a tendency to precipitate at isoelectric focusing, there are several remedies (Vesterberg, 1970, 1973a). It is also known that when problems of solubility or stability occur it is sometimes worth trying different types of dense solutes, e.g., sorbitol, ethanediol, or glycerol, to make up the density gradient. For example, in studies on cellulolytic enzymes ethanediol was found most suitable for some activities, whereas glycerol or sucrose was more suited for others (Vesterberg, 1968).

Because the additives (e.g., sucrose) are often used in a very large amount relative to the protein, it is important to exclude the influence of heavy metals and other impurities by using high-quality additives or by trying different types. During isoelectric focusing oxygen is generated at the anode. Dissolved oxygen or other products with oxidizing capability (e.g., chlorine) may spread from the anode by diffusion and interfere with proteins. The side effects of the oxidation of proteins are possible oxidation of susceptible groups in proteins (e.g., —SH or methionine) which can result in loss of activity and shift in pI that can be recorded as artifacts (Jacobs, 1973). There are ways to exclude this effect at least partly: (a) by using a short focusing time or (b) by addition of a reducing agent at the anode that takes care of the oxygen. Ascorbic acid seems to be suitable for this purpose; it seems in most cases that 0.1 g of ascorbic acid is enough for the 110-ml column (Vesterberg, 1973a). Larger quantities have been used with success (Jacobs, 1973). Another remedy that could be used to circumvent such problems is addition of thiol compounds at focusing in density gradients or in the gel technique (Vesterberg, 1970).

However, irrespective of the stabilizing effects discussed herein, problems may occur when proteins are exposed to pH values not compatible with their stability. There are general remedies for this problem also (Vesterberg, 1973a). Keep the temperature as low as possible and first of all make the time of exposure at the critical pH as short as possible. This can be done by:

(a)   introduction of the protein at a safe pH, preferably close to the pI but after the pH gradient has been established by a preceding focusing;

(b)   rapid focusing as described by Rilbe (1973b) or by working in poly-acrylamide gels where focusing can be made in about 1 hr by using an efficient cooling (Söderholm *et al.*, 1972; Vesterberg, 1973c); or

(c)   taking care of the protein as soon as possible after a separation by adjusting the pH of the fractions in question by addition of a small amount of a suitable buffer with high ionic strength.

It is well known that most metals that are bound to proteins obey an equilibrium reaction with a smaller or larger stability constant. The consequence of this is that there is usually a very small amount of free metal in solution, and as this is removed by dialysis or electrolysis more is released from the protein. In this connection it is noted that the loss of enzymatic activity for arginases seems to be of similar magnitude in dialysis and electrofocusing (Hirsch-Kolb *et al.*, 1970). The activity could afterward be increased considerably by addition of Mn^{2+} (Hirsch-Kolb *et al.*, 1970). There are other examples where much of the activity could be regained simply by activation with the proper metal (Hirsch-Kolb *et al.*, 1970; Latner *et al.*, 1970; Möllby *et al.*, 1973). In some cases preincubation of the metal is necessary; in other cases it seems sufficient to use an excess in the assay mixture. Should these methods fail, try to cut down the focusing time and the time of exposure at the pI. This is because the removal of metal is often pH and time dependent. Sometimes it is enough to dilute the protein before assay or it may be necessary to remove the CAs by some means (see Section VIII). Finally, another method is to add an excess of the important metal to the protein solution before focusing so that it is well saturated, or to add a small amount of the metal intermittently or continuously on the acidic side relative to the protein in the apparatus. Metal ions have a tendency to migrate to the cathode at the electrolysis process.

VIII.   SEPARATION OF CARRIER AMPHOLYTES FROM PROTEINS

For many purposes, e.g., detection of enzymatic activities and other bio-logical studies, it is in most cases not necessary to separate the CAs from proteins (cf. Section X). However, sometimes it is of interest to perform a separation. Most separation methods for molecules depend on physico-chemical differences. On selection of a method it is sometimes worthwhile to consider if a large number of fractions are to be treated. Since most carrier ampholytes have a molecular weight below 1000 (although for a minor percentage of the Ampholine mixture in current use, this figure may be somewhat higher according to Gasparic and Rosengren, 1974), it is natural to select methods that depend on differences in size, such as molecular sieving in gels. Experiments performed with the purpose of making a quick separation and keeping the dilution factor low have been conducted with radioactively labeled CA. These experiments have shown that it is possible to perform the separation on short columns of Sephadex gels efficiently and in less than 1 hr (Vesterberg, 1969b, 1970; Dean and Messer, 1975).

Other experiments with labeled CA have been conducted using dialysis as the separation procedure (Vesterberg, 1970). This is quite a good method for removing most of the CA. However, this is very time-consuming when it is necessary to remove even the last traces of CA, unless the dialysis procedure is done very efficiently. It may be advantageous to use a very high ionic strength in the separation to prevent possible binding of CA to proteins.

Successful separations have been performed by precipitating proteins with ammonium sulfate and then washing the precipitates. When care is taken to decrease the influence of impurities, e.g., by using high qualities of the salt, the yield of biological activities can be quite good (Nilsson et al., 1970). Other agents that can be tried for precipitation purposes are polyethylene glycol and possibly ethanol (Vesterberg, 1970). It may also be advantageous to have a high ionic strength in these experiments for the reasons already mentioned. Other procedures that may be tried are phase separations similar to procedures described by Albertsson (1960) and countercurrent procedures. The latter method may be easy to use especially for those who know the behavior of their proteins or polypeptides in such systems. This also pertains to the use of ionic exchangers for the separations (Brown and Green, 1970; Wallen and Wiman, 1972). Although the proteins and CAs in the fractions in question often have very close pI values there are still good possibilities for separation because of the differences in size, charge density, and distri-bution. Thus CA molecules usually have very small hydrophobic regions in comparison with proteins. For the reasons mentioned hydrophobic inter-action chromatography (Hjertén, 1973) could also be expected to be success-ful. The difference in size explains why separation by electrophoresis is generally very efficient. Here it can be remarked that gel electrophoresis may be a suitable procedure to check the efficiency of other separation procedures. It has been found that the CA stain very well with Amido Black. The CA will usually appear as blue spots with a higher mobility than the proteins (Dale and Latner, 1969). With isotachophoresis it has been found that CAs with the same pI as proteins usually have a higher mobility. Work is now in progress for making CAs with lower molecular weight than the ones in current use, which increases the possibilities for simpler and quicker separations.

Many substances, e.g., CA, interfere in the Lowry protein assay. However, this can be avoided (Bensadoun, 1975). The biuret procedure is somewhat better and may be used if the protein concentration is high enough to permit dilution before analysis or if a certain background can be allowed. When using protein stains after focusing in gels Bromophenol Blue (Awdeh, 1969) and also Coomassie Blue (Vesterberg, 1972; Söderholm et al., 1972) can be used without previous removal of the Ampholine. Staining for lipoproteins (Godolphin and Stinson, 1974) and glycoproteins (Graesslin et al., 1971) has also been reported to be possible without previous removal of the CAs.

IX. POSSIBLE INTERACTION WITH PROTEINS

It is well known by many chemists that weaker and stronger interactions occur between proteins and many ions in solutions. Thus binding of simple inorganic ions such as chloride and acetate has been found to occur even during electrophoresis (Tiselius and Svensson, 1940). Because protein molecules vary greatly in their charge and other surface properties, as do the CAs also (although these are more similar to each other), we could at least in a few cases expect binding forces between molecules. The question is how strong should the binding be in order to be regarded significant. If the binding of a protein and a CA of the same pI occurs, it will probably not shift the pI and can thus be disregarded for the isoelectric separation and pI determination. Here special procedures may be required for their subsequent separation, e.g., an increased ionic strength or electrophoresis (cf. Section VIII). If, on one hand, a protein and CA of different pI form a complex too stable to be broken during the isoelectric separation in question, this can cause a shift in pI that can be significant. However, if the isoelectric separation method is only used for analytical or preparative separation, as is very often the case, this phenomenon may not necessarily be unfavorable. Complex formation was reported for wool proteins (Frater, 1970), but the reaction was probably caused by precipitation at a pH incompatible with the solubility of the proteins, which is not surprising since wool proteins are well known to have a low solubility. Evidence against strong binding of CA to proteins can be found in publications by Vesterberg (1969a, 1970) and

Dean and Messer (1975).* When binding is expected there are several ways to avoid it. First an increased field strength and perhaps a prolonged focusing time can be tried. The fact that there are really strong forces present to separate molecules during isoelectric focusing is evident from experiments with Bromophenol Blue and protein mixtures containing serum albumin, which is known to bind this stain strongly. This complex is easily separable during isoelectric focusing in contrast to ordinary gel electrophoresis. Another procedure that may be tried for disintegration is the use of urea or formamide. Finally, if it is expected that a certain pH of the solution or pI of the CA is responsible for the binding, this may be avoided simply by placing the protein initially at a more favorable position between the electrodes, preferably after preformation of the pH gradient.

It was mentioned previously that the CAs since they are aminocarboxylic acids, seem to be a stabilizing and solubilizing environment for proteins (Vesterberg, 1968, 1973a). In many cases it has been found that the yield of biological activities after isoelectric focusing is as high or higher than that found after submitting the proteins to other separation procedures. This may also be explainable by the focusing, i.e., concentrating character of the method. It is known that many proteins require a certain concentration to be stable.

When a sufficient yield is not obtained it may be worthwhile to try introduction of the protein after preformation of the pH gradient; a really short focusing time (about 1 hr) (Vesterberg, 1973c; Söderholm et al., 1973; Rilbe, 1973b); addition of a certain cofactor or metal; and adjustment of pH after separation (Vesterberg, 1973c). Today when there are methods for using really short focusing times it may be worthwhile to try isoelectric focusing even in such cases where the yield was previously found low. In many cases limited success can be simply explained by the very long time first used with density gradients, e.g., 48 hr or longer.

X. BIOLOGICAL AND BIOCHEMICAL PROPERTIES

It has been found that biological activities can in most cases be assayed directly in the fractions after isoelectric separations. Should an activity be lower than expected a certain dilution may be effective, before dialysis or another procedure for removal of the CAs is tried. In some cases it has been found that addition of metal ions proper for the protein in question can increase the yield considerably, e.g., zinc ions for phosphatases (Latner et al., 1970) and hemolysins (Möllby and Wadström, 1973). In other instances it

---

* Evidence against binding as well as a procedure for separations of protein and CA has been described recently by Baumann and Chrambach (1975).

has been found that binding of protein to walls of columns or test tubes was responsible for a low regain of activity. This could be decreased by siliconization or other treatment of the columns (Hensten-Pettersen, 1974).

In some instances it is of interest to administer protein fractions to animals. Experiments on rats, mice, and rabbits have shown that CAs for these purposes could be regarded as nontoxic (Wadström et al., 1974). When larger quantities are to be injected it may be good to adjust the pH of the fractions to a more physiological one and to adjust the osmotic pressure, e.g., decrease of a high sucrose concentration, by dilution or dialysis. For those interested in cell studies it may be mentioned that the CAs have been found mild in such studies (Wadström et al., 1974), and have even been used as pH stabilizers in cell cultures (Walther and Schubert, 1974).

It seems that CAs do not interfere with immunological reactions such as immunoprecipitation and antistreptolysin assays (Leaback, 1975). Complement fixation and hemagglutination inhibition tests which are known to be critical for external factors showed no influence of CAs (Wadström et al., 1974). Should some influence in an assay system be observed it is of course possible to remove the CAs by one of the procedures mentioned in Section VIII.

XI. OTHER ASPECTS

The CAs have been found very useful in numerous applications for focusing of proteins. They have also been used as buffers in ion exchange chromatography (Leaback, 1975). This seems especially advantageous because the pI of proteins usually has some relationship to their adsorption and elution properties. In the applications reported a certain pH range of CAs was especially good, giving a high degree of purification, and in addition it was possible to submit the fractions obtained directly to isoelectric focusing without the need for removing salt ions, as is often required for protein fractions after ion exchange chromatography. The good purification obtained may partly be explained by some form of displacement chromatographic effect. Since for this principle it is usually difficult to find optimal displacing ions, it is advantageous that the CAs are a mixture of many isomers and homologs so there is a good probability that at least some molecular forms in the mixture would be suitable. The suitability of the CAs in hydrophobic interaction chromatography (Hjertén, 1973) remains to be studied.

Finally, let us comment on other ampholyte systems. The Ampholine mixture represents a significant advance concerning the number and properties of CAs in comparison with mixtures described earlier (Vesterberg, 1969b; Winogradow et al., 1973). Winogradow et al. (1973) have described a system that in principle is very similar to the one developed by Vesterberg

(1969). Although the buffer properties and some other physicochemical properties may be satisfactory for many molecules in these mixtures there are too few different molecular types and unsatisfactory relative proportions to accept these mixtures for general use. Furthermore, it would be very difficult for a biochemist to describe what he has actually been using, since the results of organic chemical reactions are generally not strictly predictable, and it may be that he has been using an unsuitable mixture for his particular protein sample. Thus, to avoid misleading observations and reports thereof, and to facilitate comparison of results on proteins (e.g., pI values) it can be stated that it is advantageous if different researchers use a CA mixture of similar composition.

REFERENCES

Albertson, P. Å. (1960). "Partition of Cell Particles and Macromolecules." Wiley, New York.
Almgren, M. (1971). *Chem. Scripta* **1**, 69–75.
Awdeh, Z. L. (1969). *Sci. Tools* **16**, 42–45.
Baumann, G., and Chrambach, A. (1975). *Anal. Biochem.* **69**, 649–651.
Besadoun, A., and Weinstein, D. (1975). *Anal. Biochem.* **70**, 241–250.
Brown, D. W., and Green, S. (1970). *Anal. Biochem.* **34**, 593–595.
Cram, D. J., and Hammond, G. S. (1964). "Organic Chemistry." McGraw-Hill, New York.
Dale, G., and Latner, A. L. (1969). *Clin. Chim. Acta* **24**, 61–68.
Davies, H. (1970). *In* "Protides of the Biological Fluids" (H. Peeters, ed.) Vol. 17, pp. 389–396. Pergamon, Oxford.
Dean, R. T., and Messer, M. (1975). *J. Chromatogr.* **105**, 353–358.
Felgenhauer, K., and Pak, S. J. (1973). *Ann. N.Y. Acad. Sci.* **209**, 147–153.
Frater, R. (1970). *J. Chromatogr.* **50**, 469–474.
Fredriksson, S., and Pettersson, S. (1974). *Acta Chem. Scand.* **B28**, 370.
Gasparic, V., and Rosengren, Å. (1974). *Sci. Tools* **21**, 1–2.
Godolphin, W. J., and Stinson, R. A. (1974). *Clin. Chim. Acta* **56**, 97–103.
Graesslin, D., Trautwein, A., and Bettendorf, G. (1971). *J. Chromatogr.* **63**, 475–477.
Haglund, H. (1971). *Methods Biochem. Anal.* **19**, 1–104.
Hensten-Pettersen, A. (1974). *Anal. Biochem.* **57**, 296–298.
Hirsch-Kolb, H., Heine, J. P., Kolb, H. J., and Greenberg, D. M. (1970). *Comp. Biochem. Physiol.* **37**, 345–359.
Hjertén, S. (1973). *J. Chromatogr.* **87**, 325–331.
Jacobs, S. (1973). *Analyst* **98**, 25–33.
Jonsson, M., and Pettersson, E. (1968). *Acta Chem. Scand.* **22**, 712–714.
Jonsson, M., Petersson, E., and Reinhammar, B. (1968). *Acta Chem. Scand.* **22**, 2135–2140.
Kolin, A. (1954). *J. Chem. Phys.* **22**, 1628–1629.
Latner, A. L., Parsons, M. E., and Skillen, A. W. (1970). *Biochem. J.* **118**, 299–302.
Leaback, D. H. (1975). *In* "Isoelectric Focusing and Isotachophoresis" (P. Righetti, ed.). Elsevier, Amsterdam.
Miner, G. D., and Heston, L. L. (1972). *Anal. Biochem.* **50**, 313–316.
Möllby, R., and Wadström, T. (1973). *Biochim. Biophys. Acta* **321**, 569–584.
Nilsson, P., Wadström, T., and Vesterberg, O. (1970). *Biochim. Biophys. Acta* **221**, 146–148.
Radola, B. (1973). *Ann. N.Y. Acad. Sci.* **209**, 127–143.

Rilbe, H. (1973a). *Ann. N.Y. Acad. Sci.* **209**, 11–22.
Rilbe, H. (1973b). *Ann. N.Y. Acad. Sci.* **209**, 80–93.
Söderholm, J., Allestam, P., and Wadström, T. (1972). *FEBS Lett.* **24**, 89–92.
Svensson, H. (1961). *Acta Chem. Scand.* **15**, 325–341.
Svensson, H. (1962). *Acta Chem. Scand.* **16**, 456–466.
Tiselius, A. (1941). *Sv. Kem. Tidskr.* **53**, 305–310.
Tiselius, A., and Svensson, H. (1940). *Trans. Faraday Soc.* **36**, 16–22.
Ui, N. (1973). *Ann. N.Y. Acad. Sci.* **209**, 198–207.
Vesterberg, O. (1968). *Sv. Kem. Tidskr.* **80**, 213–225.
Vesterberg, O. (1969a). *Sci. Tools* **16**, 24–27.
Vesterberg, O. (1969b). *Acta Chem. Scand.* **23**, 2653–2666.
Vesterberg, O. (1970). *In* "Protides of the Biological Fluids" (H. Peeters, ed.), Vol. 17, pp. 383–387. Pergamon, Oxford.
Vesterberg, O. (1971). *Methods Enzymol.* **22**, 389–412.
Vesterberg, O. (1972). *Biochim. Biophys. Acta* **257**, 11–19.
Vesterberg, O. (1973a). *Ann. N.Y. Acad. Sci.* **209**, 23–33.
Vesterberg, O. (1973b). *Acta Chem. Scand.* **27**, 2415–2420.
Vesterberg, O. (1973c). *Sci. Tools* **20**, 22–29.
Vesterberg, O. (1975). *In* "Isoelectric Focusing and Isotachophoresis" (P. Righetti, ed.), Round Table Discussion. Elsevier, Amsterdam.
Vesterberg, O., and Svensson, H. (1966). *Acta Chem. Scand.* **20**, 820–834.
Winogradow, S. N., Lowenkron, S., Andonian, H. R., Bergshaw, J., Felgenhauer, K., and Park, S. J. (1973). *Biochem. Biophys. Res. Commun.* **54**, 501–505.
Wadström, T., Möllby, R., Jeansson, S., and Wretlind, B. (1974). *Sci. Tools* **21**, 2–4.
Wallén, P., and Wiman, B. (1972). *Biochim. Biophys. Acta* **257**, 122–134.
Walther, F., and Schubert, J. C. F. (1974). *Blut* **28**, 211.

4 ISOELECTRIC FOCUSING ON POLYACRYLAMIDE GEL

Andreas Chrambach and Gerhard Baumann

Reproduction Research Branch
National Institute of Child Health and Human Development
National Institutes of Health
Bethesda, Maryland

I. INTRODUCTION: ISOELECTRIC FOCUSING IN TRANSITION

In the early 1960s pH gradient electrophoresis in a natural pH gradient, isoelectric focusing (IF) was born (Svensson, 1961), and eight years later the synthesis of carrier ampholytes was reported (Vesterberg, 1969). The success story of protein fractionation by IF has been told in four monographs (Catsimpoolas, 1973a,b; Allen and Maurer, 1974; Arbuthnott, 1974; Righetti, 1975) over the past few years. In addition, many reviews, of which only the latest will be cited for reference (Righetti and Drysdale, 1974; Catsimpoolas, 1975a), have appeared on the subject.

As this success story unfolded over the last decade, the early optimism of fractionation scientists has given way to a more critical view. The recognition of problems in IF not originally envisaged has led to attempts to improve the method. This chapter deals with this maturation process. An attempt is made to illuminate current problems in IF and to point out possible solutions in the areas of theory, mechanisms, ampholyte chemistry, and apparatus.

We will also largely restrict ourselves to IF carried out in the anticonvective medium, polyacrylamide (reviewed by Righetti and Drysdale, 1974). Isoelectric focusing in polyacrylamide (IFPA) has been nearly universally accepted as the analytical instrument of IF. This is partly because of load economy and operational simplicity. At least equally important, however, for the wide acceptance of IFPA has been (a) the enhanced zone sharpness in polyacrylamide gels compared to that obtained in sucrose density gradients, and (b) the compatibility of IFPA with protein precipitation at the isoelectric point (pI). Such isoelectric precipitation occurs as a rule at conventional protein loads used for preparative purposes. These advantages of the gel technique would make it desirable to apply IFPA, rather than IF in density gradients, on a preparative scale as well as analytically.

II. THEORY OF ISOELECTRIC FOCUSING

Although it may not seem germane to the subject of IFPA, it is necessary to consider some of the current problems in the theoretical understanding of IF in order to use and interpret IFPA data properly. The original concept of IF as an equilibrium state (Svensson, 1961) was based on sucrose density gradient experiments, which at relatively low potential gradients developed a linear pH gradient and focused proteins into bands of constant pI after one day of operation. Subsequently, band positions and pH values at the various protein zones appeared unaltered for a duration of up to two weeks (Rilbe, cited in Chrambach et al., 1973, p. 62) and only a "slight" cathodic drift ensued after that time. In view of these facts it was, and still is, perfectly justifiable to disregard minor late perturbations of the system. Considering only the initial and final states of IF, as in a thermodynamic view of the process, amphoteric molecules placed into a prefabricated pH gradient migrate electrophoretically until they come to rest at pH positions corresponding to their pI values. The original theory is not interested in the rates at which various species reach pI positions, in the late cathodic drift, or in the mechanism by which a finite current is maintained at the steady state.

Before expanding the "first approximation theory," it seems necessary to consider its semantics, for the purpose of a clearer understanding of IF. The term "equilibrium," denoting the final state of IF, is ill chosen. The term belongs to thermodynamics, and has the very specific meaning of describing the state of perfect balance between forward and backward reactions. Isoelectric focusing does not fit this concept since it involves at least the continuous transport of protons and hydroxyl ions at a rate sufficient to produce a small but finite current flow. Since there is net transport, there cannot be equilibrium. The proper term for the "final" state in IF is therefore "steady state" (Catsimpoolas, 1973b, p. 517).

Continuous optical monitoring (Catsimpoolas et al., 1975) of the separation during focusing has made it evident that all amphoteric electrolytes, carrier ampholytes or proteins approach the steady state at characteristic individual rates, and that the duration of the steady state for each species is both limited and varied. The decay of natural pH gradients (p. 81) again follows individual kinetics for each species. The concept emerging from a kinetic view of IF is therefore one of chaos. A lack of synchrony exists between the sequential stages traversed by each species at individual rates. A theoretical formulation that would adequately describe the nonsynchronized movement of the various species is evidently nearly impossible, particularly if we consider that electrophoretic migration takes place within an electrolyte milieu that differs from one point to the next in pH, field strength, temperature, etc. (Weiss *et al.*, 1974; Catsimpoolas, 1975b). Furthermore, although the migration of the various components is nonsynchronized, it is likely that at the low ionic strength used in IF intermolecular forces participate in the orderly alignment of components at the steady state and in the gradual disruption of the steady state by electrophoretic migration *en bloc* (see p. 82). Thus, association–dissociation equilibria between components may be superimposed on the formation and decay of the pH gradient.

Since a realistic theory of IF, taking into account the various kinetic and associative states, is nearly impossible, the original first-approximation theory, despite its shortcomings, remains the only workable one. What probably can be done, and should be done, is a theoretical appraisal of the migration of a representative single amphoteric species toward the steady state, of its steady state, and of its migration toward the cathode progressing with the decay of the steady state. A theoretical study, concerned only with the migration toward the steady state under a number of simplifying assumptions, has been carried out by Weiss *et al.* (1974). However, experimental validation of this study is not yet possible, since methods to determine field strength along the migration path have not been developed, and "equiconductance" preparations of ampholytes (see p. 83) are currently not available.

III. INTERPRETATION OF p*I*

The advent of IF has made the determination of p*I* exceedingly simple as compared to classical methods. The p*I* values obtained in IF are comparable to those derived from protein solutions of various ionic strengths by extrapolation to zero ionic strength (Ui, 1971; Tanford and Nozaki, 1966). However, some pitfalls do exist and should be avoided. In general, the term p*I* in IF refers to the pH region reached by the protein or ampholyte component after an arbitrary time of focusing in a gel or other anticonvective medium of arbitrary composition and at an arbitrary temperature. Such a loose use of the term p*I* makes the comparison of results between laboratories difficult. It is therefore necessary to define isoelectric points determined by IF at such a time at which they have reached a constant value, and to normalize values to standard conditions (see below). Even then, they should be designated as "apparent pI" (pI') values in contradistinction to the values determined by titration or other physicochemical methods, since they derive from a milieu of variable carrier ampholytes, variable ionic strength, etc.

How is it possible to know whether or not a pI' in IFPA has been reached? This can be done in two ways:

1. As a generally applicable method, we may carry out a focusing experiment for various durations of time, plot the pH at which the component of interest is found as a function of focusing time, and determine the plateau value of pH, using at least three reasonably spaced time points to establish the plateau (Rogol et al., 1975).

2. In specific cases where the component of interest can be recognized during focusing, be it by color (Williamson et al., 1973) or ultraviolet absorbance (Catsimpoolas, 1975b), the component of interest may be loaded throughout the gel or at both cathodic and anodic ends, and be allowed to approach a pI position from both directions. In this case, the pI' is defined as the pH at the point of coalescence of the anodically and cathodically migrating bands. It is incorrect, however, to deduce from coalescence of two colored marker proteins of particular size and mobility the time at which any other proteins might reach their isoelectric positions (see p. 88). The use of amphoteric indicator dyes as markers of the isoelectric end point has been suggested (Conway-Jacobs and Lewin, 1971). However, after attainment of their isoelectric positions, the color disappears quickly (Chrambach, unpublished data), making this approach to the determination of pH gradients impractical. However, amphoteric dyes may be useful markers for the isoelectric end point in a "nonrestrictive" gel if admixed into the gel, using the above-mentioned coalescence of the dye bands as a criterion of the isoelectric end point.

The pI' value legitimately obtained by one of these approaches is still in need of normalization to standard conditions. This normalization refers to correction of pH values measured at high viscosity to the viscosity of water, values measured at room temperature to the temperature of focusing, usually $0-4°C$, values measured in the presence of CO_2 to values obtained in its absence, and values measured at varying ampholyte concentrations to a standard concentration, e.g., 1%. From evidence with three pI ranges of ampholytes, these correction factors are appreciable (Rogol et al., 1975). If such normalization of pI' values is not carried out, it is imperative to indicate the presence and concentration of sucrose (e.g., p$I'_{30\% \, sucr}$), taking into consideration the approximate sucrose concentration in the pH fraction of interest. It is also necessary to indicate whether or not CO_2 has been removed. This appears to be significant for very dilute ampholyte solutions which arise with suspensions of gel slices in dilute KCl solution used for pH measurement. Ampholyte concentrations of 1% or more do not seem to require correction for CO_2 in the fractions, probably because of sufficient buffering capacity (Rogol et al., 1975). Correction for operative temperature, or better yet, to a standard temperature of 0–4°C, also should be indicated in the interest of interconvertibility of data.

As mentioned previously, much of the earlier fractionation work in IFPA was carried out without sufficient regard to focusing time and gel restrictiveness, i.e., without concern over whether the steady state has been reached for the species of interest. Semantic orthodoxy would require, in those cases, an indication that IF, as usually defined by its steady state, was not involved. However, in some cases, successful fractionations can be carried out without attaining the steady state. At least two reports exist in the literature where astonishing resolution was attained apparently during the transient state of IF (Park, 1973; O'Farrell, 1975), and under conditions of effective molecular sieving (i.e., using polyacrylamide gel electrophoresis with the carrier ampholytes as buffer). In one of the cases, an even distribution of protein bands along the gel was aimed at and achieved without concern about reaching the steady state. Similarly, the IF of nonamphoteric nucleic acids provides useful fractionation which cannot be isoelectric (Drysdale, 1975).

IV. MECHANISM OF ISOELECTRIC FOCUSING

The dynamics of pH gradient formation, steady state, and decay have recently been studied using ^{14}C-Ampholine (pI range 3.5–10; Catsimpoolas et al., 1975), ^{3}H-water, and highly mechanically stable (10 %T, 2 %C) gels in IFPA (Baumann and Chrambach, 1975c). Ethylenediamine (0.4%) and phosphoric acid (5%) were used as catholytes and anolytes, respectively. The vertical distance of the gel surface from the cathode was approximately 10 cm, but subsequently similar results were found when this distance was decreased to 3 mm. The temperature of IFPA was 0–4°C. Under these conditions, the pH gradient had become linear within 1 hr at 200 V. However, the distribution of Ampholine through the gel was nonuniform from the beginning: Acidic Ampholine components predominated; the concentration of all other species decreased in proportion to pH. During the steady state (3–9 hr) a progressive decrease of Ampholine concentration in the neutral portion of the pH gradient was observed, accompanied by the appearance of a basic Ampholine peak. The previously noted pH "plateau phenomenon" (Finlayson and Chrambach, 1971) is probably the result of this occurrence. While this basic peak formed, a joint displacement of all amphoteric species toward the cathode was observed. The formation of a basic ampholyte peak during the steady state and the migration of the ampholytes toward the cathode are incompatible with a true equilibrium state. In agreement with previous findings, both in IFPA (Chrambach *et al.*, 1973) and in IF (Catsimpoolas, 1973a), pH gradients were found to be unstable with time. This instability was not due to a "progressive accumulation of water at the neutral gel region," as had been previously postulated (Chrambach *et al.*, 1973), but to a progressive accumulation or formation of acidic and basic ampholytes and corresponding depletion of neutral species, and to predominantly cathodic, but also anodic, electrophoresis of ampholytes. This ampholyte migration was shown to be at best partially electroendosmotic: endosmosis accounted for not more than 10% of the net Ampholine transport. The different components of Ampholine appeared to be displaced at the same rate. It is therefore possible that they migrated as a "steady-state stack" (see below) or were held together by noncovalent intermolecular forces. As the acidic Ampholine peak traversed the gel toward the cathode, acid followed. This migration continued until the entire gel had acidified and was depleted of ampholytes. The conclusion that Ampholine movement is electrophoretic seems inescapable, although it is based on indirect evidence. Assuming the correctness of this conclusion, it is necessary to postulate that focusing is not truly isoelectric, although for practical purposes it may be assumed to approximate the isoelectric condition during the steady state. (Recent evidence for this postulate is the formation of a natural pH gradient with nonamphoteric buffers (Nguyen and Chrambach, 1976).

The notion that IF is a form of steady-state stacking (isotachophoresis, ITP) has been recently advanced (Chrambach *et al.*, 1975). The one obvious difference is that in ITP one or several moving boundaries are generated that move continuously in the field, while the isoelectric zones in IF are stationary in the field (as a first approximation). The question, discussed later, is whether this difference rules out a unitary mechanism between ITP and IF. Isotachophoresis arises under conditions of electrolyte orientation such that (a) a more rapidly migrating electrolyte precedes a less rapidly migrating one in the direction of electrophoretic migration, and (b) the two are separated by a stationary boundary. The electrolyte orientation in order of pI—and therefore of mobility—among carrier ampholytes is achieved through their enclosure into a "cage" of acid and base that would tend to reverse the net charge of the amphoteric species on contact. The "stationary boundary" between adjacent ampholytes is provided by the anticonvectant medium (gel or density gradient).

Another way of viewing IF would be as a form of "cascade stacking." Here, as suggested by Jovin (Catsimpoolas, 1973b, p. 522), buffer constituents varying in pK values and mobilities would be allowed to set up a succession (cascade) of moving boundaries into which proteins of the same mobilities would "fit." Within such a buffer cascade a pH gradient forms. When the entire stack is placed into a "pH cage" (see p. 82), a stationary alignment of boundaries is obtained, which again gives rise to a stable gradient of pH, temperature, and voltage, i.e., to IF (Nguyen *et al.*, in preparation).

V. SYNTHETIC CARRIER AMPHOLYTES

Widespread use of Ampholine over the years has also yielded insight into its limitations and shortcomings. The most serious of these is the unevenness of conductance along the pH gradient, conductance being much higher in the acidic and basic gel regions than in the neutral zone (Frater, 1970). This leads to differential Joule heating at the initial "high current stage" and also possibly later, where no significant heating is expected. Also, the unevenness of field strength makes it impossible to determine mobility in the gradient, or many other significant physicochemical parameters that IF could otherwise yield (Catsimpoolas, 1975b).

Currently available Ampholine does not contain an uninterrupted pI spectrum of ampholyte species: Conductivity gaps have been observed in the pH gradient (Stenman, 1975). Also, it has been difficult, until recently, to produce very acidic and very basic ampholyte species, so that ampholytes were "fortified" with basic and acidic amino acids, as evidenced by amino acid analysis (Hayes and Wellner, 1969) and by optical rotatory dispersion measurements (Righetti *et al.*, 1975b).

Another problem with ampholytes has been the difficulty in separating them completely from proteins. Some components appear of such large molecular size that they are nondialyzable and elute in gel fitration together with a protein of 22,000 MW (molecular weight) (Baumann and Chrambach, (1975a). It appears from these data that the previous gel filtration data (Gasparic and Rosengren, 1974) yielded too low a molecular weight distribution, possibly through retardation of the polyethylene glycol molecular weight standards, or their relatively higher hydration compared with that of the Ampholine components.

Spectroscopic measurements have found evidence for nitrogen heterocyclic structures which the synthesis of ampholytes should not produce (Righetti *et al.*, 1975b). It is not known whether these cyclic elements are noxious to the stability of ampholytes, the formation and stability of pH gradients, etc., and whether they are unavoidable side reaction products of polymerization.

All these deficiencies of ampholytes, as well as economic motives, have led to attempts to fractionate ampholytes and recombine fractions to yield a mixture with the desired properties, or to synthesize amphoteric molecular species with properties superior to those of current ampholytes.

Fractionation of ampholytes by ion exchange chromatography (Brown *et al.*, 1975), IF and ITP (Everaerts, 1975) is possible. Isoelectric focusing is used to produce commercial ampholytes of various pI ranges, and is potentially capable of yielding fractions that can be recombined to yield mixtures with equiconductance properties. Fractionation by ITP is potentially important because of the gram-preparative capability of this method (Baumann and Chrambach, 1975d) and the fact that analytical ITP of ampholytes has yielded several hundred fractions (Everaerts, 1975), which is 10 times more than any other method has yielded to date. Ion exchange chromatography has been successfully applied to the fractionation of synthetic carrier ampholytes (Brown *et al.*, 1975). However, since this fractionation is based in part on hydrophobic interactions and adsorption on the resin, this method seems less suited to yield fractions differing in pI.

Ampholytes with different amine-to-carboxylic-acid ratios (Righetti *et al.*, 1975a) and ampholytes with different types of substituted amines (Righetti *et al.*, 1975a; Vinogradov *et al.*, 1973) have been synthesized. Of potential interest in view of the deficiencies in neutral ampholyte species is the synthesis of amphoteric compounds containing functional groups with neutral pK, such as imidazole. For the synthesis of more acidic ampholytes, the partial substitution of carboxylic by sulfonic groups is significant (Pogacar and Jarecki, 1974). It is to be expected that with increasing diversity of available ampholytes, equiconductive, optically inactive, low-molecular-weight amphoteric compounds can be developed.

We have alluded to the problem of separating ampholytes from protein on the basis of molecular size, because of the presence of large-molecular-weight species. However, it appears possible quantitatively to separate ampholytes from protein by polyacrylamide gel electrophoresis (PAGE) (Baumann and Chrambach, 1975a), by ion exchange chromatography on a mixed bed ion exchange resin (Brown and Green, 1970; Baumann and Chrambach, 1975b), or by repetitive ammonium sulfate precipitation of the protein (Li and Li, 1973). Surprisingly, the much dreaded interaction between ampholytes and protein does not seem to take place, as found by the use of labeled ampholytes and labeled protein (Baumann and Chrambach, 1975a,b). Polyacrylamide gel electrophoresis provides a sensitive assay for the presence of residual ampholytes in protein preparations: ampholytes are recognized at the proper gel concentrations as a poorly fixing material forming a halo around the position of the stack (Baumann and Chrambach, 1975a).

Detection of carrier ampholytes in the presence of protein has been feasible, taking advantage of their refractivity under dark-field, sideways illumination (Righetti *et al.*, 1975a), of the caramelization reaction of acidic ampholytes (Felgenhauer and Pak, 1973), and specific fixation (Svendsen and Chrambach, 1975). However, it remains unknown to what degree each method detects groups of carrier ampholytes, and what the distribution of amphoteric species for each group is. At this time, a discrepancy of at least one order of magnitude appears to exist between the number of ampholyte species detected conductimetrically on ITP (Everaerts, 1975) and that detected by staining methods.

VI. APPARATUS: ANALYTICAL AND PREPARATIVE

Both gel tube and gel slab apparatus have been used for analytical IFPA. Although the protagonists of slab apparatus have suggested that tube apparatus for IFPA has been made obsolete by slab apparatus (Vesterberg, 1973; Wadstrøm, 1973, p. 410), it seems that both are needed in IFPA for particular purposes.

Tube apparatus is preferable for determining the optimal focusing time, the optimal nonrestrictive gel concentration, the kinetics of pH gradient formation, the degree of resolution at the steady state as a function of time, and the optimal pI range of ampholytes to be used. All of these experiments can be done by withdrawing individual tubes from a single electrolyte reservoir during one IFPA run, while each condition would have to be studied on individual slabs in separate experiments, or by slab-slicing techniques which are not sufficiently developed.

It is equally clear that once an optimal gel concentration, focusing time, and ampholyte pI range are defined, it is advantageous to carry out IFPA of many samples on a single gel, i.e., on a slab apparatus.

Vertical slab apparatus gives rise to difficulties in the mode of sample application. When plastic slot formers are used, their removal from mechanically labile nonrestrictive gels is not always possible without collapse of the slot or damage to the gel. When solid partitions between sample wells are used, band distortion easily occurs in the segmented electric field when sample volumes, viscosity, or ionic strengths are not identical in the wells. The advantage of vertical over horizontal slabs is the possibility of applying samples directly into the wells with minimal adsorptive losses.

Horizontal slab apparatus has the advantage of being applicable to mechanically labile gels, such as polyacrylamide gels of less than 4 %T, 2–5 %C, or more than 10 %C, polyacrylamide–agarose mixtures, or agarose gels. Sample application by insertion of filter paper strips into the gel appears optimal (Smyth and Wadstrøm, 1975). Prefabricated, dried, and rehydrated polyacrylamide–agarose gel slabs may be of potential use in horizontal slab apparatus.

More important than the geometry of the apparatus used is uniform and and efficient heat dissipation which permits the use of high voltages, gel uniformity, and field symmetry required for obtaining straight bands. Uniformity of the electric field is not usually obtained in application of buffer wicks to the gel. Heating is of relatively minor importance in steady-state IF as compared to PAGE. However, heat dissipation is important during the early transient state of IF. It can possibly be significant at later stages in regions of high voltage gradients (see p. 83). The requirement for optimal heat dissipation characteristics of the apparatus does not subside with the advent of constant power supplies (Schaffer and Johnson, 1973; Søderholm and Wadstrøm, 1974), although the danger of "cooking" the protein is diminished as long as moderate power is applied. Since resolution in IFPA improves with increasing voltage, power should in fact not be kept "reasonable" but should be maximized. Resolution of hemoglobin has been shown to improve dramatically in IFPA when high voltages were applied at the terminal stages of fractionation (R. A. Allen, personal communication). Such maximization again makes it necessary to build apparatus with the best possible heat dissipation characteristics. Thin and heat-ductile walls between the gel and the coolant (glass rather than plastic), the cooling of the gel around its entire circumference rather than only along one face, and possible substitution of liquid coolant flow with electronic methods of heat transfer (e.g., utilizing the Peltier effect) may help in achieving better heat dissipation. Most available apparatus fail in one or all of these criteria; however, noncommercial prototypes exist for both tube (Chrambach *et al.*, 1975) and horizontal slab (Johansson, 1972) apparatus which represent attempts at optimization.

Two-dimensional apparatus has recently gained new importance for IFPA in view of the astonishing degree of resolution obtained on unfractionated, radioactively labeled *Escherichia coli* extracts by combination in two dimensions of IFPA with sodium dodecyl sulfate (SDS)–PAGE (O'Farrell, 1975).

Preparative apparatus for IF in density gradients has been widely used since the beginnings of IF. It has been pointed out that density gradient IF is inherently burdened with isoelectric precipitation. In addition, loss of resolution occurs during elution, unless it proceeds without turbulence and is carried out in the electric field (Svendsen, 1969). Preparative IFPA apparatus is nonexistent to date, except in the form of tube gels of 18 mm diameter (Finlayson and Chrambach, 1971; Chrambach *et al.*, 1975) which can be sliced and eluted (Rogol *et al.*, 1975). Preparative IFPA remains of interest, however, as resolution and load capacity are much improved in gels. A suitable procedure for preparative IFPA has recently been tested (McCormick *et al.*, 1976). It takes advantage of the fact that any protein zone at isoelectric point pI can be electrophoretically isolated by replacing one of the electrolytes with an ampholyte of first pI + ΔpI, then, when the component of interest has migrated to the end of the gel, with ampholyte of pI − ΔpI (or the reverse). Under these conditions, the component of interest can be eluted into a small, circulating volume of ampholyte (with pI + ΔpI or pI − ΔpI), using conventional elution PAGE preparative apparatus (Kapadia *et al.*, 1974; Chrambach *et al.*, 1975). Alternatively, elution can be effected by pH gradient elution using buffers instead of catholyte or anolyte, after the steady state has been reached.

Preparative IF apparatus using anticonvective media possessing flow properties (e.g., Sephadex) is available in the form of the Sephadex flat bed apparatus (Radola, 1973) and the continuous transverse field apparatus (Fawcett, 1973). The first has the advantage of instrument simplicity and commercial availability. However, it depends on a completely homogeneous Sephadex bed with reproducible and specific moisture content to give straight bands across the bed. To date, no simple foolproof way to achieve such a bed is available. The apparatus for continuous elution again excels through its simplicity of design. However, it consumes more ampholytes than most workers can currently afford. (This is obviously not a serious long-range restriction in view of the synthetic approaches just mentioned and the possibility of regenerating the ampholytes.) An inherent design problem is the absence of a detection system that would allow one to judge when the pH gradient has linearized in the direction perpendicular to continuous flow at the bottom of the cell. This problem could be solved by providing "optics" (Catsimpoolas *et al.*, 1975) or use of indicators, but possibly only at considerable sacrifice of simplicity and cost.

A preparative IF apparatus compatible with isoelectric precipitation, consisting of an array of multiple V-shaped compartments, was devised early in the development of IF (Valmet, 1969). Here, the isoelectric precipitates of protein are collected at the bottom of the V-shaped compartments containing ampholytes at the various pH levels. Relatively high voltage is required. This model has yet to be built in dimensions small enough for the usual laboratory, so that it can be more widely tested. It cannot be predicted at this time whether this or preparative IFPA apparatus will be the successor to current apparatus for preparative density gradient IF, which is incompatible with the isoelectric precipitation of proteins.

VII. GEL

The postulate that the gel used for IFPA should be nonrestrictive to the migration of the protein under study has to be viewed in a relative sense since all gels are necessarily restrictive to all molecules.

The application of excessively restrictive gel concentrations to IFPA has been the single most decisive error in much of the past work in IFPA. It should be recognized that under conditions of molecular sieving proteins may not reach their isoelectric positions in the time allotted by the finite life span of the pH gradient. Thus, the position of the protein after an arbitrary time of migration in a restrictive gel may not be isoelectric. To avoid this problem, two approaches have been taken.

1.   Colored proteins, or otherwise detectable proteins, have been applied at both ends of the pH gradient, and the isoelectric end point has been defined by the coalescence of zones progressing toward the pI position from both polarities (Catsimpoolas, 1975b; Williamson et al., 1973). Using this approach, the problem of gel restrictiveness also arises since the two bands may merge only after the decay of the pH gradient has set in, with concomitant loss of resolution. This method presents even more problems when applied to noncolored proteins, i.e., in the usual case. Here, the colored protein of size X has been used to gauge the time and gel concentration required for band coalescence of an uncolored protein of size Y. This is evidently invalid unless the colored marker and the unknown protein are both unrestricted in their migration by the gel and exhibit similar mobilities.

2.   It has been suggested to define a nonrestrictive gel arbitrarily as one which allows the protein to migrate in PAGE at an extreme of pH with a relative mobility $(R_f) = 1.0$ (Chrambach et al., 1973). This means that at this particular gel concentration the mobility of the protein equals that of the moving boundary in multiphasic buffer systems (Chrambach et al., 1975). This definition of a tolerable gel concentration for IFPA does provide relatively wide-pore gels in practice, although it is theoretically unsatisfactory since it neglects the dependence of stacking on the rate of propagation of the moving boundary in question, or, in terms of the Jovin theory (Chrambach et al., 1975), the value of RM(1, ZETA) or RM(1, PI). If this propagation rate is low, proteins will remain stacked even at relatively high gel concentration; if it is high, molecular sieving in a restrictive gel will retard the migration of proteins below the propagation rate of the moving boundary. The described "stacking test" at any arbitrary value of RM (1, ZETA) or RM (1, PI) is not necessarily representative of the conditions prevailing in IFPA. However, it is possible to compare the relative nonrestrictiveness of two gels accurately by measuring the retardation coefficients K_R, which are independent of boundary displacement rates. In practice, the finding of a tolerable gel concentration imposes the burden of constructing a Ferguson plot for the protein of interest prior to subjecting it to IFPA. This approach also involves the incorrect assumption that an open-pore, labile gel applicable in PAGE would be equally applicable to IFPA.

Wall adherence in IFPA is more precarious than in PAGE for two reasons.

1.   Ampholyte concentrations and pH vary at different points along the gel, causing the degree of hydration (swelling) to vary correspondingly. A nonuniformly swollen gel is evidently less firmly adhering to glass walls than the uniform gel used in PAGE.

2.   Some ampholytes act as amine initiators in the polymerization reaction of acrylamide, causing a decrease in the average polymer chain length and greater mechanical lability of the gel. Highly cross-linked gels provide considerably larger pore sizes than the usual (2–5%) degree of cross-linking (Rodbard et al., 1972), but their mechanical properties are inferior.

A compromise between the conflicting requirements of open pore size and mechanical stability therefore has to be found. It seems useful and appropriate to select a generally applicable nonrestrictive gel concentration for all molecules within a certain size range. Thus, for all molecules of less than 0.5 million MW, a 5 %T, 15 %C_{DATD} gel seems applicable (Baumann and Chrambach, 1976). The cross-linking agent N,N'-diallytardiamide (DATD) has been chosen for this purpose to provide the necessary mechanical stability. In addition, it provides transparent gels; gels highly cross-linked with Bis are opaque. For IFPA fractionation of molecules in excess of 0.5 million MW, Bis should be used as the cross-linking agent since it provides larger effective pore sizes than DATD. Polyacrylamide gels with up to 50% cross-linking can be prepared (Rodbard et al., 1972), so that IFPA can be carried out without practical limitation due to molecular size. Possibly, it may even lend itself to particle separation.

VIII.   PROTEIN FRACTIONATION

Staining of proteins in the presence of ampholytes used to be very difficult since all stains customarily applied to proteins (Coomassie Brilliant Blue G- or R-250, Amido Black, Fast Green, Coomassie Violet) also react with carrier ampholytes (Catsimpoolas, 1973b, p. 152). Recently, a selective "no background" staining method for protein in the presence of ampholytes using perchloric acid fixative has been described (Reisner et al., 1975).

SDS proteins can be fractionated by IF, since the detergent appears quantitatively removed during IF (Friesen et al., 1971). Isoelectric focusing in polyacrylamide is also fully compatible with the presence of the nonionic detergent Triton X-100 and therefore applicable to most water-insoluble proteins (Miller and Elgin, 1974).

A.   Protein Solubilization

In preparative applications of IFPA, or in the case where the protein fractions are to be analyzed by activity, it is frequently not possible to diffuse the isoelectric protein from gel slice at a convenient rate. It is then necessary to solubilize the isoelectrically precipitated protein by exposure to high pH (approximately 10.5) at 0°C for 5 to 20 min. The solution may then be neutralized (Kaplan et al., 1972; Ben-David et al, 1974).

Solubilization of isoelectrically precipitated protein is also effected by SDS. This is particularly useful in two-dimensional IFPA–(SDS–PAGE) (Stegemann, 1969; O'Farrell, 1975).

ACKNOWLEDGMENTS

Ms. Terri Sellner expertly entered text and revisions into the IBM Wylbur text-editing system.

REFERENCES

Allen, R. C., and Maurer, H. R. (eds.) (1974). "Electrophoresis and Isoelectric Focusing on Polyacrylamide Gel." Gruyter, Berlin.

Arbuthnott, J. P. (ed.) (1974). "Isoelectric Focusing." Butterworths, London.

Baumann, G., and Chrambach, A. (1975a). *Anal. Biochem.* **64**, 530.

Baumann, G., and Chrambach, A. (1975b). *Anal. Biochem.* **69**, 649.

Baumann, G., and Chrambach, A. (1975c). *In* Righetti (1975).

Baumann, G., and Chrambach, A. (1975d). *Fed. Proc.* **34**, 685; (1976) *Proc. Nat. Acad. Sci. USA* **73**, 732.

Baumann, G., and Chrambach, A. (1976). *Anal. Biochem.* **70**, 32.

Ben-David, M., Becker, R., Rodbard, D., and Chrambach, A. (1974). *Endocrinol. Res. Commun.* **1**, 211.

Brown, W. D., and Green, S. (1970). *Anal. Biochem.* **34**, 591.

Brown, R. K., Lull, J. M., Bagshaw, J. C., Lowenkron, S., and Vinogradov, S. N. (1975). *Fed. Proc.* **34**, 685; (1976) *Anal. Biochem.* **71**, 12.

Catsimpoolas, N. (1973a). *Ann. N.Y. Acad. Sci.* **209**, 65.

Catsimpoolas, N. (ed.) (1973b). *Ann. N.Y. Acad. Sci.* **209**, 1–529.

Catsimpoolas, N. (1975a). *Separ. Sci.* **10**, 55.

Catsimpoolas, N. (1975b). *In* Righetti (1975).

Catsimpoolas, N., Griffith, A. L., Williams, J. M., Chrambach, A., and Rodbard, D. (1975). *Anal Biochem.* **69**, 372.

Chrambach, A., Doerr, P., Finlayson, G. R., Miles, L. E. M., Sherlins, R., and Rodbard, D. (1973). *Ann. N.Y. Acad. Sci.* **209**, 44.

Chrambach, A., Jovin, T. M., Svendsen, P. J., and Rodbard, D. (1975). *In* "Methods of Protein Separation," Vol. 1 (N. Catsimpoolas, ed.) p. 27. Plenum, New York.

Conway-Jacobs, A., and Lewin, L. M. (1971). *Anal. Biochem.* **43**, 394.

Drysdale, J. W. (1975). *In* Righetti (1975).

Everaerts, F. (1975). *In* Righetti (1975).

Fawcett, J. S. (1973). *Ann. N.Y. Acad. Sci.* **209**, 112.

Felgenhauer, K., and Pak, S. J. (1973). *Ann. N.Y. Acad. Sci.* **209**, 147.

Finlayson, R., and Chrambach, A. (1971). *Anal. Biochem.* **40**, 292.

Frater, R. (1970). *Anal. Biochem.* **38**, 536.

Friesen, A. D., Jamieson, J. C., and Ashton, F. E. (1971). *Anal. Biochem.* **41**, 149.

Gasparic, V., and Rosengren, A. (1974). *In* "Isoelectric Focusing" (U. P. Arbuthnott and A. Beeley, eds.), p. 178. Butterworths, London.

Hayes, M. B., and Wellner, D. (1969). *J. Biol. Chem.* **244**, 6636.

Johansson, B. G. (1972). *Scand. J. Clin. Lab. Invest. Suppl. 124* **29**, 7.

Kapadia, G., Chrambach, A., and Rodbard, D. (1974). *In* "Electrophoresis and Isoelectric Focusing on Polyacrylamide Gel" (R. C. Allen and H. R. Maurer, eds.), p. 115. Gruyter, Berlin.

Kaplan, G. N., Maffezzoli, R. D., and Chrambach, A. (1972). *J. Clin. Endocrinol. Metab.* **34**, 370.

Li, Y. -T., and Li, S. C. (1973). *Ann. N.Y. Acad. Sci.* **209**, 187.

McCormick, A., Miles, L. E. M., and Chrambach, A. (1976). *Anal. Biochem.* **74** (in press).

Miller, D. W., and Elgin, S. C. R. (1974). *Anal. Biochem.* **60**, 142.

Nguyen, N. Y., and Chrambach, A. (1976). *Anal. Biochem.* (In press).

Nguyen, N. Y., Rodbard, D., Svendsen, P. J., and Chrambach, A. In preparation.

O'Farrell, P. H. (1975). *J. Biol. Chem.* **250**, 4007.

Park, C. M. (1973). *Ann. N.Y. Acad. Sci.* **209**, 237.

Pogacar, P., and Jarecki, J. (1974). *In* "Electrophoresis and Isoelectric Focusing on Poly-acrylamide Gel" (R. C. Allen and H. R. Maurer, eds.), p. 153. Gruyter, Berlin.

Radola, B. J. (1973). *Biochim. Biophys. Acta* **295**, 412; *Ann. N.Y. Acad. Sci.* **209**, 127.

Reisner, A. H., Nemes, P., and Bucholtz, C. (1975). *Anal. Biochem.* **64**, 509.

Righetti, P. G., and Drysdale, J. W. (1974). *J. Chromatogr.* **98**, 271.

Righetti, P. G. (ed.) (1975). "Recent Progress in Isoelectric Focusing and Isotachophoresis". Elsevier, Amsterdam.

Righetti, P. G., Pagani, M., and Gianazza, E. (1975a). *J. Chromatogr.* **109**, 341.

Righetti, P. G., Righetti, A. R. B., and Galante, E. (1975b). *Anal. Biochem.* **63**, 423.

Rogol, A. D., Ben-David, M., Sheats, R., Rodbard, R., and Chrambach, A. (1975). *Endocrinol. Res. Comm.* **2**, 379.

Rodbard, D., Levitov, C., and Chrambach, A. (1972). *Separ. Sci.* **7**, 705.

Schaffer, H. E., and Johnson, F. M. (1973). *Anal. Biochem.* **51**, 577.

Smyth, C. J., and Wadström, T. (1975). *Anal. Biochem.* **65**, 137.

Söderholm, J., and Wadström, T. (1974). *In* "Isoelectric Focusing" (U. P. Arbuthnott and J. A. Beeley, ed.), p. 132. Butterworths, London.

Stegemann, H. (1969). *Z. Anal. Chem.* **350**, 917.

Stenman, U. H. (1975). *In* Righetti (1975).

Svendsen, P. J. (1970). *Protides Biol. Fluids Proc. Colloq.* **17**, 413.

Svendsen, P. J., and Chrambach, A. (1975). In preparation.

Svensson, H. (1961). *Acta Chem. Scand.* **15**, 325.

Tanford, C., and Nozaki, Y. (1966). *J. Biol. Chem.* **241**, 2832.

Ui, N. (1971). *Biochim. Biophys. Acta* **229**, 567.

Valmet, E. (1969). *Protides Biol. Fluids Proc. Colloq.* **16**, 401.

Vesterberg, O. (1969). *Acta Chem. Scand.* **23**, 2653.

Vesterberg, O. (1973). *Ann. N.Y. Acad. Sci.* **209**, 145.

Vinogradov, S. N., Lowenkron, S., Andomian, M. R., and Bagshaw, J. C. (1973). *Biochem. Biophys. Res. Commun.* **54**, 50.

Wadström, T. (1973). *Ann. N.Y. Acad. Sci.* **209**, 405.

Weiss, G. H., Catsimpoolas, N., and Rodbard, D. (1974). *Arch. Biochem. Biophys.* **163**, 106.

Williamson, A. R., Salaman, M. R., and Kreth, H. R. (1973). *Ann. N.Y. Acad. Sci.* **209**, 210.

5 ISOELECTRIC FOCUSING/ ELECTROPHORESIS IN GELS

Colin W. Wrigley

Wheat Research Unit, CSIRO
North Ryde, New South Wales, Australia

I. INTRODUCTION

The technique of gel isoelectric focusing was conceived from the principles of isoelectric focusing in density gradients, yet it was actually born out of the techniques of gel electrophoresis, and for its nuture and development has relied heavily on the procedures and apparatus, the skills and know-how originally devised for gel electrophoresis. It has thus been natural that so many reports of protein fractionation by gel isoelectric focusing compare its resolving capacity with that previously obtained by gel electrophoresis.

In many cases superior resolution and greater heterogeneity have been reported for gel isoelectric focusing as compared with gel electrophoresis. Nevertheless, gel electrofocusing in no way replaces gel electrophoresis, since the principles of fractionation are different in each case. Both are important methods of protein fractionation; both are valuable tests of protein homogeneity.

However, separate application of each of the two methods singly to a protein sample, as very often reported, does not realize the full potential of using them both. With little extra effort, the two can be used in combination to provide a more meaningful two-dimensional separation using techniques and apparatus already developed for two-dimensional electrophoresis.

For example, to report that a preparation was resolved into, say, 10 zones by isoelectric focusing and six by electrophoresis does not provide nearly as much information as showing, by combining the methods, that electrophoretic zones 1, 3, and 6 remained homogeneous on electrofocusing and that each of the remaining zones was further resolved by isoelectric focusing. Additional information can also be obtained by specific staining methods.

In designing an experiment using gel electrofocusing combined with gel electrophoresis, there is a wide range of conditions that might be used, especially for the electrophoretic stage. This chapter sets out to explain and illustrate this range of combinations. To assist in choosing the best conditions, the purpose of the experiment should be clear. It is thus appropriate to indicate first the type of information that might be sought.

II. AIMS OF GEL ELECTROFOCUSING/GEL ELECTROPHORESIS

The two-dimensional pattern shown in Fig. 1 is offered as an example of gel electrofocusing/gel electrophoresis to illustrate the purposes that might be achieved by the method.

A. Investigation of Protein Heterogeneity

The terms homogeneous and heterogeneous have meaning only in relation to the method used for analysis. As pointed out in Section I, these terms have more meaning, especially for components of a mixture, when related to a combination of gel isoelectric focusing and gel electrophoresis than to the separate methods. The combined procedure is useful in checking the purity of a zone which appears slightly diffuse, without definite separation into two zones, by fractionation in one dimension. For another case, subsequent

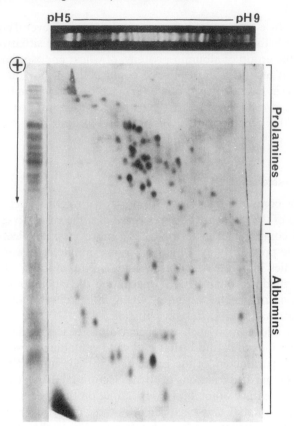

Fig. 1. Combined gel isoelectric focusing and electrophoresis of proteins from wheat endosperm, by the method of Wrigley and Shepherd (1973). The pattern obtained by electrofocusing in the first dimension appears across the top. The strip at left shows the pattern given by gel electrophoresis (pH 3) alone. Starch gel was used for the second stage (electrophoresis) in preference to polyacrylamide, since starch gel has proved better for resolving this particular mixture of proteins. In general, however, polyacrylamide is preferred.

electrophoresis might be useful in improving spatial separation of zones focused close together, without actually altering the assessment of heterogeneity. Even if homogeneity is demonstrated for a preparation by so critical a test as this, differences in composition especially due to uncharged amino acids could still remain undetected.

As shown in Fig. 1, gel electrophoresis of the prolamines (gliadin proteins) of wheat grain endosperm reveals about 15 components, and gel isoelectric focusing about 20. Originally it was supposed that roughly the same 15–20 proteins were resolved in both cases, but two-dimensional separation revealed over twice this number of components (Wrigley, 1970a). Various zones separated by either method alone were made up of several components which were resolvable by the other procedure. A similar situation held for the water-soluble group of albumin proteins of wheat endosperm, as shown in the lower half of Fig. 1. No evidence could be found for the possibility that the increase in resolution was due to artifact formation. On the contrary, the fact that the two-dimensional patterns proved useful in providing information about the genetic control of protein synthesis, suggests that the separation is real (Wrigley and Shepherd, 1973).

B. Protein Characterization

A reason for combining gel electrophoresis with gel isoelectric focusing in the early use of the latter method has been, in some cases, to characterize electrofocused components in terms of the more familiar electrophoretic mobilities of the previously used method. For example, this was partly the aim of the studies illustrated in Figs. 1 and 7. The addition of isoelectric separation for gliadin proteins (Fig. 1) permitted the elucidation of electrophoretic analysis of genetic variants, when interpretation of the single-dimensional data had been extremely difficult.

However, in addition to such empirical aims, it is possible to choose specific combinations of methods that will yield valuable information about physicochemical parameters of the proteins involved, without the need for actually isolating purified fractions of them. Section IV provides recommendations about the choice of methods, depending on the specific information required. Parameters that may be studied by these analytical means include isoelectric point, molecular weight, subunit size, charge characteristics, and immunological properties.

C. Developing a Strategy for Protein Purification

The task of devising a sequence of steps for purifying a particular protein from a crude mixture has tended to be approached in a largely empirical or almost intuitive manner. However, the development of analytical methods for characterizing proteins in mixtures, together with identification by specific staining methods, offers the possibility of planning the strategy for purification on the basis of the analytically determined characteristics of the protein of interest and of the other components in the mixture.

An attractive aspect of gel isoelectric focusing is that the results it provides analytically (it is not wholly suitable for preparative separation) are almost directly related to results that might be expected from the corresponding preparative procedure, isoelectric focusing in a density gradient. This is shown by comparisons reported, for example, by Wellner and Hayes (1973).

On the other hand, the adaptation of gel electrophoresis for true preparative fractionation has met with only limited success.

However, characterization of the proteins involved, as suggested in the preceding section, provides information that is valuable in choosing the best sequence of preparative methods. A suitable combination of gel electro-focusing and electrophoretic methods is also useful for assessing progress during purification, for thus checking the choice of the next step, and for establishing final "purity."

Robinson and Leaback (1974) suggest the following sequence of steps for a generalized approach to protein purification: Check pH stability of the protein of interest. Collect isoelectric and size information about it and associated proteins by two-dimensional analyses. If appropriate to the mix-ture, purify initially on the basis of size by gel filtration. Then, using infor-mation about isoelectric points, establish conditions for further purification of the protein of interest using ion exchange cellulose. Purify further using density-stabilized isoelectric focusing and gel filtration, again taking advan-tage of the known characteristics of the proteins remaining after each stage of purification.

To refer again to Fig. 1, the reason for initially combining the two methods in this case was largely to determine which gliadin components might be best suited to purification. A system of ion exchange chromatography had been established which fractionated components roughly in the sequence of their mobilities on gel electrophoresis (Wrigley, 1965). Examination of the two-dimensional pattern thus indicated several prominent components, well separated from others, as being the easiest to purify. The two-dimensional display also gave a lead to the best way to combine ion exchange chroma-tography with preparative isoelectric focusing for this purpose.

D.   Displaying Protein Composition

Combined gel electrofocusing/electrophoresis has proved quite valuable for analytical screening of protein composition for complex mixtures. For this purpose the tube/slab procedure (Section III.C) is most suitable. Such two-dimensional patterns of protein composition have been referred to as "protein maps," since they provide a ready display or "fingerprint" of the presence or absence of individual components of a characterized mixture. Protein mapping of serum proteins for diagnostic purposes is an excellent example of this application (Dale and Latner, 1969). Its use in diagnosis has also been extended to other body fluids (Latner, 1973). Protein mapping of gliadin composition (Fig. 1) was used to screen a series of wheat lines, each with specific chromosome deficiencies so that the chromosomes responsible for gliadin protein synthesis could be identified.

E. Removal of Carrier Ampholytes

An important bonus, often available when gel electrophoresis is used after isoelectric focusing, is the separation of fractionated proteins from the carrier ampholytes used in the isoelectric focusing step. If this can be arranged, it avoids the need for tedious removal of ampholytes before applying common protein-staining methods or for resorting to other staining methods with which ampholytes do not interfere. However, achieving the removal of ampholytes is not necessarily the main point of using electrophoresis after isoelectric focusing.

Success in removing ampholytes depends largely on the choice of pH conditions. For the experiment shown in Fig. 1, pH 5–9 ampholytes were used for focusing. On subsequent electrophoresis at pH 3, these ampholytes were all positively charged and they quickly moved cathodically (down the gel) well ahead of the proteins. Some of the slowest moving ampholytes (pH 5) can be seen at the bottom of the gel as a strongly stained area. Such ampholyte removal is generally possible by performing electrophoresis at a pH outside the focusing pH range. Alternatively, pore gradient electrophoresis (Section IV.D) can be used to remove carrier ampholytes; protein migration becomes restricted by the decreasing pore size, thus accentuating the faster mobility of the ampholytes which pass freely through the gel pores.

III. GENERAL TECHNIQUES AND APPARATUS

Details of specific combinations of gel isoelectric focusing with the many available forms of gel electrophoresis are provided and illustrated in Section IV. These need to be prefaced with comments on general aspects of procedure and technique.

A. The Isoelectric Focusing Step

Chapters 4 and 6 provide a full discussion of procedures for isoelectric focusing in either tubes or slabs. For completeness within this chapter, a brief description is given of a procedure found to be satisfactory and convenient. It may be applied to gel tubes or slabs. In the latter case, it is particularly important to guard against leaks down the side of the gel and to maintain constant conditions across the gel (thickness, temperature, and conductivity).

On the other hand, there are too many different types of electrophoretic procedures for detailed instructions to be given in each case. In general, the electrophoretic methods are well established, and specific references are offered in Section IV as a source of technical details.

1. Gel Preparation

The following formula makes sufficient gel for about 20 × 70 × 3 mm gel tubes or one gel slab (70 × 70 × 3 mm). Final concentrations are 2% for carrier ampholytes* and 6% for acrylamide monomer.

Gel mixture (Mix just before use)
Acrylamide solution   4.0 ml
Carrier ampholytes (Ampholine, 40%)   1.0 ml
Riboflavin (10 mg/100 ml) (fresh solution)   1.0 ml
Water, to make a final volume of   20.0 ml
Acrylamide solution (Protect from light, renew monthly)
Acrylamide   30 g
N, N'-methylene bisacrylamide   1 g
Water, to make a final volume of   100 ml

Remove dissolved air from the gel mixture by applying a vacuum for a few minutes. Fill gel tubes to within about 5 mm of the top and overlayer with water. For the gel slab, overlayer with water if applying only one sample right across or if it is to accept a gel rod for focusing in the second dimension. Otherwise, insert a slot former or similar device to form pockets in the top of the gel for loading a range of samples. Photopolymerize by standing the gels about 10 cm from a fluorescent light. This should be complete after 20 to 30 min at 20°C.

Alternatively, gels may be chemically polymerized (without illumination) by replacing the riboflavin solution with a fresh solution of potassium persulfate (10 mg/ml). Polymerization should be complete about 30 min after mixing, at 20°C. This gel is usually stronger than a photopolymerized gel, but residual persulfate has been known to modify proteins during electrofocusing. Overnight storage or prefocusing is helpful in removing persulfate.

2. Sample Application

Mount gel tubes or slab in the apparatus. Cover the gel surface with a protective layer (2–5 mm) containing 2% carrier ampholytes and 5% sucrose. This protects the sample (made up in 10% sucrose and layered between the upper layer and the gel) from contact with the extreme pH value of the electrode solution.

Alternatively, the protein sample may be incorporated in the gel mixture before polymerization. This permits the use of very dilute protein samples, but prevents the examination of a number of different samples with a slab gel.

---

* From LKB-Produkter AB, Stockholm, Sweden.

The amount of protein to be loaded depends on the sensitivity of the means of detection, the size of the gel, and the number of zones it produces. About 20 μg per zone is a common loading for a 70 × 5 mm tube gel. If such a gel is to be used for further fractionation by gel electrophoresis, a higher loading will be needed because of spreading of zones in the second dimension and the possibility of resolving the sample into further components.

3.   Focusing of Proteins

Choose the orientation of electrodes according to the method of sample loading and its pH stability. Carefully pour in the electrode solutions: 0.4% sulfuric acid for the anode compartment and 0.4% ethanolamine for the cathode.

Turn on the current, and gradually raise the voltage (as conductivity falls) to about 60 V/cm of gel length. If samples are to be applied to the top of gel tubes or into pockets on top of a gel slab, this may be done after prefocusing for about 30 min. Under these conditions, 90–120 min is generally adequate for protein focusing.

It is most important to ensure that focusing is complete. This can be done by focusing separate gels for different times to check that zones have focused to a reasonably stable pattern. Alternatively, it is possible to see the positions of strong zones during focusing, as refractile zones. For a slab gel, the refractile zones may be seen as shadows cast onto a screen by shining a small strong light through the gel.

4.   Protein Detection

After the run, zones of most proteins can be visualized in the gel as white bands of precipitation by several minutes' immersion in 10% trichloroacetic acid. Extensive washing in trichloroacetic acid solution is needed to remove carrier ampholytes before using conventional protein staining procedures. However, direct staining methods have been developed for use after electrofocusing. The following procedure (Vesterberg, 1971) has given satisfactory results with a range of proteins. Immerse gels in the stain for 15 min at 60°C (in a fume hood) or for 1 hr at 20°C. Destain by soaking in several changes of ethanol–acetic acid–water (3:1:6).

Direct staining solution

Coomassie Brilliant Blue G-250	100 mg
Trichloroacetic acid	10 g
Sulfosalicylic acid	3 g
Methanol	25 ml
Water, to make a final volume of	100 ml

If focused gels are to be used for subsequent electrophoresis, this is best done soon afterward, so that zone sharpness is not lost by diffusion. Alternatively, polyacrylamide gels should be frozen if they are to be stored overnight or longer, before electrophoresis in the second dimension. Preliminary soaking of the gel before fractionation in the second dimension is generally unnecessary.

B.   Sequence of Using the Two Methods

In most combinations of gel isoelectric focusing with electrophoresis, the focusing step has been used first. This has partly been because some workers have felt more at ease with electrofocusing in tubes and electrophoresis in slabs than with the reverse combination. While gel electrofocusing in slabs has since become more popular, there are still good general reasons for retaining this sequence. Initial fractionation by focusing permits the use of dilute samples, not suitable for gel electrophoresis, and provides well-sharpened zones for the second stage of fractionation. Furthermore, use of electrophoresis as the second step offers the possibility of removing carrier ampholytes at that stage, thus simplifying the staining procedure.

On the other hand, there are no reasons why this sequence may not be reversed, if such is more convenient for a particular experiment. For example, Maurer and Allen (1972) have obtained very good fractionation of serum proteins by using gel electrofocusing after cellulose acetate electrophoresis (see Fig. 6).

C.   Whole Tube Applied to Slab Gel

The clearest picture of two-dimensional separation is provided by fractionation in a gel tube initially, followed by fractionation in a rectangular slab of gel. The tube is placed across the slab so that the directions of fractionation in the two steps are at right angles.

1.   Apparatus

The initial stage of fractionation is generally performed in apparatus of the type described by Davis (1964) for disc electrophoresis. Instead of a gel tube, a strip cut from a gel slab may be used, as described by Dewar and Latner (1970).

The techniques and apparatus needed for the second stage are those already established for two-dimensional electrophoresis. Most equipment designed for electrophoresis in a rectangular gel is suitable. Two basic designs are illustrated in Fig. 2.

The more popular design (Fig. 2a) involves laying the gel rod on the top edge of the slab. The apparatus of Akroyd (1967) is very simple and easy to

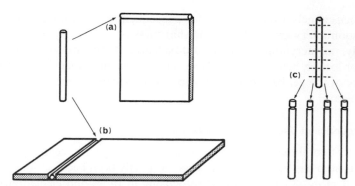

Fig. 2. Techniques for combined gel isoelectric focusing/electrophoresis. (a) and (b) Two basic designs for using a tube gel with a gel slab for two-dimensional display. (c) Use of tube gels throughout. The tube gel from the initial separation is cut into sections, each of which is applied to another gel tube for the second stage.

construct. Dale and Latner (1969) describe apparatus specifically designed for gel electrofocusing/electrophoresis. The single-slab apparatus of Margolis and Kenrick (1968) is suitable for detecting zones during fractionation by shining light through (see Section III.A.3). The design shown in Fig. 3 (Mets and Bogorad, 1974) provides for the partial concentration of the zones from the one-dimensional gel as they pass into the gel slab.

The second type of arrangement (Fig. 2b) is often used for electrophoresis in a range of media (starch gel, agarose, granulated gel layers) as well as polyacrylamide. The apparatus design is basically that of Smithies (1959)—a flat trough to contain the gel layer, preferably with a cooling tank beneath. The tube gel lies across the slab in a narrow slot formed either by cutting the gel or during the casting of the gel by a strip of plastic attached to the mold.

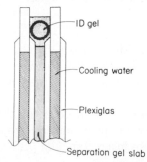

Fig. 3. Cross section of apparatus set up for the second stage of two-dimensional fractionation. The tube gel from the first stage (1D gel) appears at top in end view, above the gel slab in which the second stage of separation will take place. [From Mets and Bogorad (1974).]

2.   Technique

For all types of apparatus, it is important that a thin one-dimensional gel be used, so that starting zones are sharp for the second stage. In addition, there should be close contact between the gel tube and the surface of the slab.

For the design in Fig. 2a, the gel tube should fit the gap above the slab neatly, to ensure efficient removal of proteins into the second gel. Removal of air bubbles is important. If electrofocusing is used in the second dimension, the electrophoretic gel rod may be placed onto the top of the Ampholine gel under a protective layer of Ampholine–sucrose solution (see Section III.A.2). Alternatively, the tube may be set in Ampholine gel.

The design of Fig. 2b was used for the separation shown in Fig. 1. A 100 × 3 mm electrofocused gel was laid in a 2.5-mm-wide slot cast in position across a 4-mm-thick gel slab. To ensure electrical continuity, the slot was first filled with liquid before inserting the gel rod. Air bubbles were removed as the slot was covered with plastic.

3.   Position of Tube in Slab

If electrophoresis is applied after isoelectric focusing, care should be taken in choosing the position of the gel tube in the slab with respect to electrode polarity. The focused gel should be inserted across the cathodic end of the electrophoretic slab if the pH of electrophoresis is above the pH range of isoelectric focusing. Conversely, proteins will be positively charged if the pH of electrophoresis is below the pH range of focusing: The focused gel is placed near the anode and the proteins move cathodically into the electrophoretic gel. The focused gel must be placed across the middle of the electrophoretic gel slab if the pH of electrophoresis is within the focusing pH range since some proteins will migrate towards the anode and some toward the negative electrode.

D.   Tube Sections Applied to Tube Gels

An approximation to two-dimensional fractionation in the form just described may be achieved using tube gels throughout as shown in Fig. 2c. Disc electrophoresis equipment is used for both stages (Davis, 1964). The gel from the initial step of fractionation is cut across its length into sections. Each section is placed on top of a fresh tube of gel (same diameter or slightly larger) for the second stage of fractionation. (An example of this procedure appears in Fig. 4.)

Compared with the tube–slab method, this procedure gives a fragmented picture of two-dimensional separation. It is best suited to examination of simple mixtures of well-separated zones. If it is not possible to detect the positions of zones in the initial tube before sectioning, there is the danger

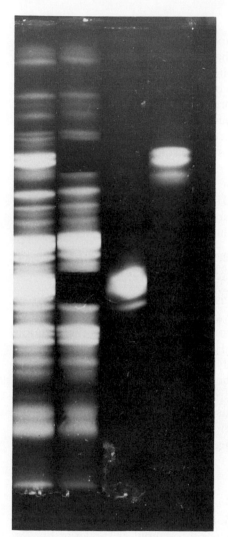

Fig. 4. Use of the sectioned-tube/tube method to check reproducibility of isoelectric focusing (Wrigley, 1970b). The gel on the left shows the whole pattern of zones obtained by isoelectric focusing of wheat gliadin proteins in the range pH 5 (top) to 9. From a duplicate gel (second from left), two sections were cut before fixing it in 5% trichloroacetic acid. Each section was applied to the top of a fresh gel (gels at right) for a rerun under similar conditions. The zones cut from the original gel focused to positions equivalent to the gaps left in it.

of cutting through the center of one zone and of thus mistakenly interpreting it as containing two components. Any detection method used to guard against this possibility must not be harmful to protein structure. Ultraviolet scanning and some forms of enzyme staining are suitable.

This procedure has been used effectively by Smith et al. (1971), who located zones of alkaline phosphatase activity on an electrofocused gel, cut the zones out, and placed each on top of an electrophoretic tube gel for further fractionation. The technique also worked well in reverse (for electrophoresis followed by electrofocusing) and thus provided a check on possible artifact formation.

Catsimpoolas (1970) has reported the use of this general approach for immunoelectrophoresis following isoelectric focusing in a gel rod. In this case, the gel sections from the focusing stage were broken into small pieces to facilitate the removal of protein during the second step.

In another variation (Nader et al., 1974), proteins were eluted from the sections of the first-stage gel, and were applied as solutions for the second stage. Generally actual elution of protein is not necessary. If it is, a preparative procedure might be preferred for the first phase of fractionation.

E. Combination with Preparative Methods

In certain circumstances, either gel isoelectric focusing or gel electrophoresis might be less suitable than an equivalent preparative method. For example, analytical information might be required about isoelectric fractions. In such cases, an electrophoretic method has been used to examine portions of protein fractions separated by isoelectric focusing in a density gradient (Robbins and Summaria, 1973).

A "two-dimensional display of molecular size–electric charge" is obtained with the procedure of Leaback and Robinson (1974) by applying analytical gel isoelectric focusing to fractions separated by preparative gel filtration. The latter method is preferred to electrophoretic methods for size determination with native proteins. This procedure may give a fragmented picture of two-dimensional separation, as does the sectioned-tube method (Section III.D), but it has proved valuable in developing a strategy for protein isolation (Robinson and Leaback, 1974).

Gel-stabilized layers, described in Chapter 6, offer a suitable medium for the second dimension of fractionation, either for electrophoresis or for isoelectric focusing. This medium offers the considerable advantage that after fractionation, protein zones may be eluted from it more readily than from polyacrylamide gel.

IV. SPECIFIC COMBINATIONS

A. Gel Isoelectric Focusing in Both Dimensions

1. A Test of Reproducibility

For any new method that produces improved resolution, the possibility must be investigated that the heterogeneity it reveals might not be real, but is due to the formation of artifacts by the procedure itself. One of a set of recommendations (Ressler, 1973) for testing for artifacts is that isolated components be rerun under similar conditions to check for reproducibility.

Two-dimensional fractionation is a convenient means of performing this check on artifact formation for gel isoelectric focusing. The protein sample in question is focused in one dimension and the resulting gel strip is used for a repeat focusing at right angles to the first experiment. Zones should focus in a straight line diagonally across the gel.

If isoelectric focusing in gel tubes is being checked, it is more satisfactory to use the sectioned-tube procedure (Section III. D). In this way the conditions of the initial experiment can be more readily repeated in the second stage. A more rigorous test of reproducibility is provided by changing the position of application for the second stage.

Figure 4 shows the results of such a check on some components of a complex mixture (Wrigley, 1970b). Whereas in this case zones refocused reproducibly, anomalous behavior on repeated isoelectric focusing has been reported by Frater (1970), Wrigley (1970b), and Felgenhauer *et al.* (1972).

2. Study of Protein Structure

Useful information about the chemical structure of sample proteins may be obtained by repeating an initial focusing run, with a change in conditions in the second stage designed to affect protein structure or conformation. This new approach offers a convenient procedure for pursuing studies, such as those of Salaman and Williamson (1971) and Ui (1971a), on the isoelectric behavior of proteins in the native and denatured states. Gel isoelectric focusing would first be applied to the native proteins, followed by focusing in a second gel containing a denaturing agent such as urea.

The additional study of subunit structure might require elution of proteins after the first stage or substitution of preparative isoelectric focusing, so that disulfide bonds might be reduced and alkylated. Alternatively, disulfide bonds might be broken within the first-stage gel by incorporating a charged thiol compound, such as 2-mercaptoethylammonium chloride (Jones *et al.*, 1972) or thioglycolate, in the second stage of separation. The charge of such a compound would carry it into the first-stage gel and produce more effective reduction than introduction of an uncharged molecule by diffusion.

A similar two-dimensional technique might also prove useful for studies on protein conformation, such as those described by Ui (1971b, 1973), involving the focusing of an enzyme in the presence and absence of an inhibitor, and involving investigation of carbohydrate interaction with concanavalin A. A discussion of factors contributing to isoelectric heterogeneity by Williamson *et al.* (1973) provides a useful introduction to studies of this type.

B. Gel Isoelectric Focusing Combined with
Gel Electrophoresis

This combination of methods indicates the isoelectric points of protein components and their charge–size relationships at the pH of electrophoresis. Separate information about charge and size requires a series of electrophoretic analyses at different gel concentrations (Kingsbury and Masters, 1970; Rodbard and Chrambach, 1971). This information might be more readily obtained by using cellulose acetate electrophoresis and either sodium dodecyl sulfate (SDS) or pore gradient gel electrophoresis.

Combined gel electrofocusing and electrophoresis has produced very good resolution in fractionating a variety of proteins. In most cases, a slab gel has been used for the second dimension (Section III.C), with electrofocusing preceding electrophoresis. Examples of this application include potato proteins (Macko and Stegemann, 1969), L-amino acid oxidase (Hayes and Wellner, 1969), and wheat endosperm proteins (Wrigley, 1970a; Wrigley and Shepherd, 1973) (see Fig. 1). Stegemann (1970) has described a procedure for extracting protein components for subsequent examination following "protein mapping."

The combination has also been successfully applied to serum proteins. The resulting two-dimensional display has proved more characteristic for interpretation in diagnostic work (Latner, 1973) than a lineup of bands resolved by one method only. The diagram of Fig. 5 represents the pattern

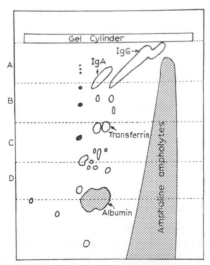

Fig. 5.   Diagram of the fractionation of proteins from normal serum by gel isoelectric focusing (pH 3–10) followed by polyacrylamide gel electrophoresis (pH 8.9). [From Dale and Latner (1969), who give detailed instruction on the procedure.]

obtained by gel isoelectric focusing/electrophoresis of only 5 μl of serum. Comparisons between this pattern for normal serum and those for pathological sera show notable differences.

The heavily staining strip down the right-hand side of the pattern is due to carrier ampholytes stained by the protein dye. These ampholytes represent the more basic of the pH 3–10 mixture used for electrofocusing. In this case, no attempt was made to remove them before staining, since previous experiments had shown that they did not hide any protein zones.

Serum proteins have also been fractionated two-dimensionally by a combination of isoelectric focusing and disc gel electrophoresis by Domschke et al. (1970).

The procedure illustrated in Fig. 5 for serum has also been applied by Latner (1973) to proteins from cerebrospinal fluid, urine, gastric juice, bile, and tissue extracts, using either direct protein staining or methods for specific detection of enzyme activity. Further details of these studies are given by Dale et al. (1970) and Fossard et al. (1970).

C. Gel Isoelectric Focusing Combined with Cellulose Acetate Electrophoresis

The combination of isoelectric fractionation with electrophoresis on cellulose acetate offers the possibility of obtaining information on protein charge properties, provided allowance is made for electroendosmosis in the electrophoretic stage. In applications of this method, electrofocusing has been used as the second stage with the cellulose acetate strip inserted into a slit cut in the Ampholine gel (Leaback et al., 1970; Maurer and Allen, 1972).

Figure 6 shows the pattern for human plasma proteins obtained by this combination of methods. At the top is the pattern of bands separated by cellulose acetate electrophoresis alone. On the left are the patterns for gamma globulins (extreme left) and for whole plasma fractionated by isoelectric focusing alone. Clearly, isoelectric focusing has made the greater contribution to resolution, but the electrophoretic step has neatly separated groups of proteins to make interpretation easier after focusing.

D. Gel Isoelectric Focusing Combined with Pore Gradient Electrophoresis

Pore gradient electrophoresis is a method of fractionating proteins in the native state according to molecular size (Margolis and Kenrick, 1968; Slater, 1969). It involves electrophoresis of proteins through a gel in which pore size decreases progressively because of an increase in gel concentration. Under these conditions, migration rate decreases until the protein molecules virtually stop in the region of the gel where pore size matches molecular

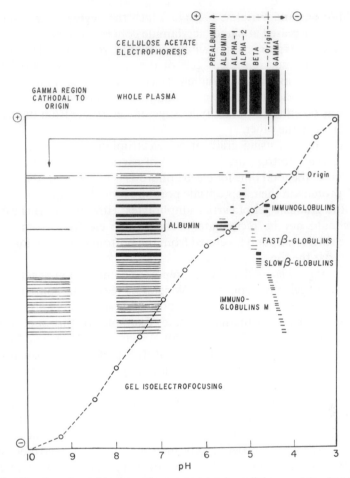

Fig. 6. Fractionation of human plasma proteins by cellulose acetate electrophoresis (one-dimension pattern, top right), followed by isoelectric focusing in 6% polyacrylamide gel. Electrofocusing patterns for gamma globulins and whole plasma appear on the left, and the two-dimensional pattern is on the right. [From Maurer and Allen (1972).]

size for each protein. Thus, an estimate of molecular weight can be obtained by comparison with the behavior of standard proteins, since charge differences between proteins are minimized.

This empirical method of determining molecular weight is only satisfactory in the "asymptotic case" where migration is almost halted (Rodbard *et al.*, 1971). In the original procedure, this requirement was met for proteins of 100,000 daltons or more, but smaller proteins were not completely retained. Retention of smaller proteins has been considerably improved in a new type of gel (Margolis and Wrigley, 1975) in which the degree of cross-linking is considerably increased at higher acrylamide concentrations.

This new gel type is commercially available, ready for electrophoresis, as Gradipore brand.* Gradient gels of the same dimensions (82 × 82 × 2.8 mm) are also available.† Gradient electrophoresis equipment, available from either of these sources, can be adapted to slab gel electrofocusing if great care is taken to prevent any leakage of electrolyte from one electrode compartment to the other.

An advantage of using gradient gel electrophoresis after gel isoelectric focusing is that electrophoretic removal of carrier ampholytes is generally possible, since the total time for electrophoresis is not critical; the time taken to "jam" proteins at their appropriate pore sizes should be more than enough to clear the gradient gel of carrier ampholytes. However, in this case, electrolyte should not be recirculated between the electrode compartments, otherwise carrier ampholytes eluted from the bottom of the gel, for example, will reenter at the top.

A gradient gel was used in the first reported combination of gel isoelectric focusing with gel electrophoresis (Wrigley, 1968). The combination has proved valuable for serum proteins because of its suitability for resolving the large serum proteins in the native state. Examples of this application are reported by Kenrick and Margolis (1970), Wright *et al.* (1973), and Emes *et al.* (1975). Wright *et al.* (1973) concluded that they preferred electrophoresis, in a homogeneous gel, to gel isoelectric focusing for the first stage of fractionation.

Figure 7 shows a series of two-dimensional separations on gradient gels following different times of isoelectric focusing. The various components of the mixture of serum proteins are labeled. Comparison of the patterns shows how the proteins have progressively migrated into the electrofocusing gel after application at its alkaline end (now the left side of the two-dimensional pattern). Focusing appears to have been complete after about 3 hr and to have remained approximately stable for an additional 3 hr.

E. Gel Isoelectric Focusing Combined with SDS Gel Electrophoresis

Another electrophoretic method for investigating protein size involves treating proteins with the anionic detergent sodium dodecyl sulfate (SDS), largely to mask charge differences between proteins (Shapiro *et al.*, 1967). The treatment has also usually involved rupture of disulfide bonds so that the method is mainly intended as an estimation of molecular weight of

---

* Gradient Pty. Ltd., Lane Cove, 2066, Australia, or Ortec Inc., Oak Ridge, Tennessee.
† Pharmacia Fine Chemicals A.B., Uppsala, Sweden.

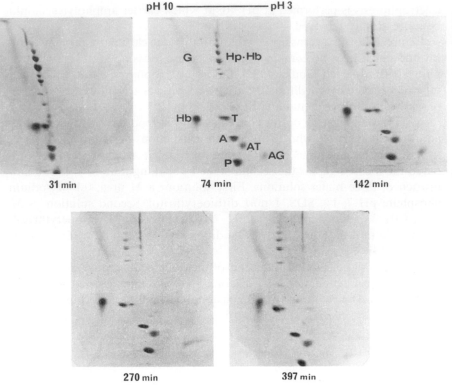

Fig. 7. Serum proteins separated by gel electrofocusing in the first dimension (for the times indicated at 120 V), followed by pore gradient electrophoresis at pH 9.3 for 20 hr at 80 V in a gel slab (4.5% gel at top, progressively increasing to 26% at the bottom). Identity of proteins: Hp · Hb, haptoglobin–hemoglobin complex; Hb, free hemoglobin; T, transferrin; A, albumin; AT, α_1-antitrypsin; P, prealbumin; G, immunoglobulin IgG; AG, α_1 acid glycoprotein. [From Kenrick and Margolis (1970), who give full experimental details.]

protein subunits (individual polypeptide chains), and is useful for investigating protein quaternary structure.

Since the method involves protein denaturation, pore gradient electrophoresis should be used instead, if enzyme staining is to be used to locate components or if native oligomeric proteins are studied. On the other hand, SDS electrophoresis is satisfactory for virtually the whole protein size range, and thus would probably be preferred to pore gradient electrophoresis for proteins smaller than, say, 30,000 daltons.

For actual performance of SDS electrophoresis, the methods of Shapiro *et al.* (1967) or Weber and Osborn (1969) are recommended. These involve anodic electrophoresis at pH 7. The method of Panyim and Chalkley (1971) might offer some advantages for combination with isoelectric focusing, since electrophoresis is performed at pH 10, at which carrier ampholytes should be more readily removed.

Figure 8 shows patterns obtained for nonhistone chromatin proteins from rat and chick liver by combined gel isoelectric focusing and SDS electrophoresis (Barret and Gould, 1973). Before electrophoresis, the focused gel tube was soaked overnight in 10% trichloroacetic acid, followed by a series of washes in buffers containing SDS.

A similar application of this combination was reported by MacGillivray and Rickwood (1974), who gave detailed instructions on the methods used, including a discontinuous buffer system for SDS electrophoresis. They prepared the focused gel for the second stage by shaking it at 40°C for 30 min in each of three buffer solutions. First solution: 8 M urea, 0.1 M sodium phosphate pH 7, 1% SDS, 1 mM dithioerythritol. Second solution: 8 M urea, 0.01 M sodium phosphate pH 7, 1% SDS, 1 mM dithioerythritol. Third solution: 8 M urea, 0.01 M sodium phosphate, 0.1 % SDS, 1 mM dithioerythritol.

The potential of this combination of methods is illustrated by its application to the fractionation of *Escherichia coli* proteins labeled with ^{14}C-amino acids. The resulting two-dimensional autoradiograph (Fig. 9) shows

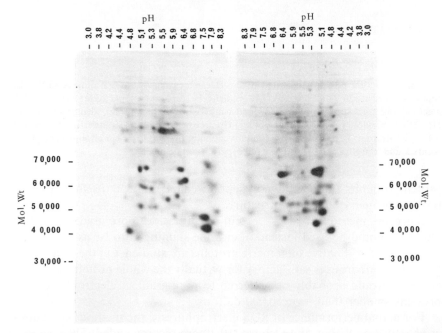

Fig. 8. Gel isoelectric focusing/SDS electrophoresis of chromatin nonhistone proteins from rat (left) and chicken. Protein loadings were 560 and 680 μg, respectively. [From Barret and Gould (1973).]

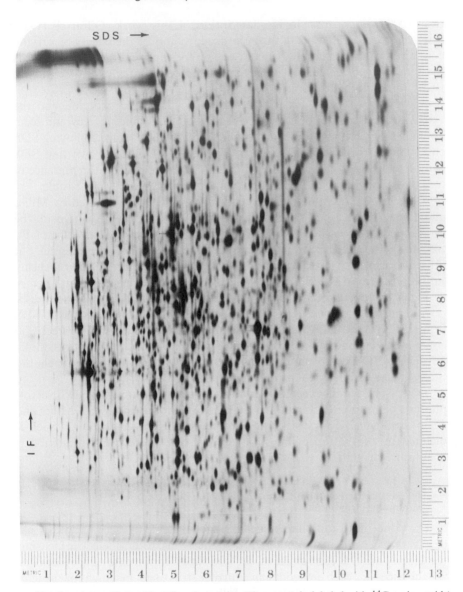

Fig. 9.   Autoradiograph of *E. coli* proteins (10 μg protein labeled with ^{14}C-amino acids) after fractionation by gel isoelectric focusing followed by SDS electrophoresis in an acrylamide gel (gradient of 9.25–14.4% gel). [From O'Farrell (1975).]

over 1000 components. In his description of this work, O'Farrell (1975) provides considerable detail of the procedure and of the effects of varying conditions of fractionation. A major reason for the great resolving power is the sensitivity of detection. A protein which constitutes as little as about $10^{-4}\%$ of the total protein can be detected.

In view of the strong interaction between SDS and proteins, and the extensive modification of charge properties by SDS binding, it would seem unlikely that normal isoelectric focusing could be performed following SDS electrophoresis. However, Weber and Kuter (1971) have shown that SDS can be removed from proteins using an anion exchanger in the presence of 6 M urea. Partial restoration of enzymatic activity was even possible.

The effect of SDS on gel isoelectric focusing has been studied by Miller and Elgin (1974). They showed that a satisfactory pH gradient could be established even with 250 μg SDS applied to the gel. The presence of 1% SDS in the sample did not alter the focused pattern for serum albumin, but other proteins such as hemoglobin gave very different patterns after focusing in the presence of 1% SDS. Some success was even achieved in an attempt to focus protein directly from a gel slice following SDS electrophoresis, relying on the presence of 8 M urea in the focusing gel to dissociate SDS from the protein. However, their recommendation was that correct and complete electrofocusing information can probably be obtained on samples purified by SDS electrophoresis only if the samples are first eluted from the SDS gel and dialyzed against 8 M urea.

Thus, isoelectric focusing is best used in combination with SDS electrophoresis before the SDS treatment step. However, the findings of Miller and Elgin are available for using the combination in reverse order if such is dictated by the nature of the problem, e.g., the use of SDS to solubilize membrane proteins.

F. Gel Isoelectric Focusing Combined with Immunoelectrophoresis

The concept of combining isoelectric focusing with electrophoresis has been very successful in immunological applications. After isoelectric focusing, the gel rod may be used intact or cut into slices. In the case of the sectioned gel, the slices are each embedded in a further strip of gel in which the proteins are fractionated electrophoretically, followed by immunochemical analysis (Catsimpoolas, 1970).

The technique applied to the whole focused gel tube is an extension of the crossed electrophoresis method of Laurell (1965). Electrophoresis is applied to the gel strip perpendicular to the direction of isoelectric focusing, and proteins migrate into an antibody-containing gel in which antigen–antibody complexes form. At the end of the experiment, the precipitation line formed at the leading edge of the migrating protein zone may be used to quantitate the various protein components. The technique has been effectively applied to urinary proteins (Rotbol, 1970), salivary amylases (Skude and Jeppsson, 1972), and human serum proteins (Jirka and Blanicky, 1973).

V. CONCLUSION

Further combinations of isoelectric focusing with electrophoretic techniques are possible and are likely to appear as development in these fields continues. For example, gel isotachophoresis (Griffith *et al.*, 1973) offers yet another element of variation to complement and extend currently popular methods.

Possibly, in the combinations of methods described, we are approaching the limits of resolution for detecting differences between proteins in the number of amino acids and the composition of charged amino acids. However, any estimate of heterogeneity or of genetic variation of amino acid content in proteins must underestimate the situation several-fold since differences due to uncharged amino acids are presumably largely neglected using electrophoresis and isoelectric focusing.

Methods based on different principles are needed for further major advances. Affinity chromatography and associated principles might provide a source of new applications. Already a marriage with electrophoresis has been attempted by Bog-Hansen (1973). His method, called "crossed immuno-affinoelectrophoresis," combines the principles of electrophoresis with those of biospecific interaction to yield a prediction of the results of affinity chromatography experiments. Presumably a similar match with isoelectric focusing might be made.

The question of whether any such novel method, or one of the combination techniques of this chapter, or an older tried and tested procedure is best suited to resolving proteins of a particular mixture, depends partly on the properties of the components and on the way the methods are used, and also on that element of art that is still an undeniable part of our analytical science.

REFERENCES

Akroyd, P. (1967). *Anal. Biochem.* **19**, 399–410.
Barret, T., and Gould, H. J. (1973). *Biochim. Biophys. Acta* **294**, 165–170.
Bog-Hansen, T. C. (1973). *Anal. Biochem.* **56**, 480–488.
Catsimpoolas, N. (1970). *Clin. Chim. Acta* **27**, 365–366.
Dale, G., and Latner, A. L. (1969). *Clin. Chim. Acta* **24**, 61–68.
Dale, G., Latner, A. L., and Muckle, T. J. (1970). *J. Clin. Pathol.* **23**, 35–38.
Davis, B. J. (1964). *Ann. N.Y. Acad. Sci.* **121**, 404–427.
Dewar, J. H., and Latner, A. L. (1970). *Clin. Chim. Acta* **28**, 149–152.

Domschke, W., Seyds, W., and Domagk, G. F. (1970). *Z. Klin. Chem. Klin. Biochem.* **8**, 319.

Emes, A. V., Latner, A. L., and Martin, J. A. (1975). *Clin. Chim. Acta* **64**, 69–78.

Felgenhauer, K., Graesslin, D., and Huismans, B. D. (1972). *In* "Protides of the Biological Fluids" (H. Peeters, ed.) (*19th Colloq., 1971*), pp. 575–578. Pergamon, Oxford.

Fossard, C., Dale, G., and Latner, A. L. (1970). *J. Clin. Pathol.* **23**, 586–589.

Frater, R. (1970). *J. Chromatogr.* **50**, 469–474.

Griffith, A., Catsimpoolas, N., and Kenney, J. (1973). *Ann. N.Y. Acad. Sci.* **209**, 457–469.

Hayes, M. B., and Wellner, D. (1969). *J. Biol. Chem.* **244**, 6636–6643.

Jirka, M., and Blanicky, P. (1973). *Biochim. Biophys. Acta* **295**, 1–7.

Jones, J. M., Creeth, J. M., and Kekwick, R. A. (1972). *Biochem. J.* **127**, 187–197.

Kenrick, K. G., and Margolis, J. (1970). *Anal. Biochem.* **33**, 204–207.

Kingsbury, N., and Masters, C. J. (1970). *Anal. Biochem.* **36**, 144–158.

Latner, A. L. (1973). *Ann. N.Y. Acad. Sci.* **209**, 281–298.

Laurell, C. B. (1965). *Anal. Biochem.* **10**, 358–361.

Leaback, D. H., and Robinson, H. K. (1974). *FEBS Lett.* **40**, 192–195.

Leaback, D. H., Rutter, A. C., and Walker, P. G. (1970). *In* "Protides of the Biological Fluids" (H. Peeters, ed.) (*17th Colloq., 1969*), pp. 423–426. Pergamon, Oxford.

MacGillivray, A. J., and Rickwood, D. (1974). *Eur. J. Biochem.* **41**, 181–190.

Macko, V., and Stegemann, H. (1969). *Hoppe-Seyler's Z. Physiol. Chem.* **350**, 917–919.

Margolis, J., and Kenrick, K. G. (1968). *Anal. Biochem.* **25**, 347–362.

Margolis, J., and Wrigley, C. W. (1975). *J. Chromatogr.* **106**, 204–209.

Maurer, H. R., and Allen, R. C. (1972). *Clin. Chim. Acta* **40**, 359–370.

Mets, L. J., and Bogorad, L. (1974). *Anal. Biochem.* **57**, 200–210.

Miller, D. W., and Elgin, S. C. R. (1974). *Anal. Biochem.* **60**, 142–148.

Nader, H. B., McDuffie, N. M., and Dietrich, C. P. (1974). *Biochem. Biophys. Res. Commun.* **57**, 488–493.

O'Farrell, P. H. (1975). *J. Biol. Chem.* **250**, 4007–4021.

Panyim, S., and Chalkley, R. (1971). *J. Biol. Chem.* **246**, 7557–7560.

Ressler, N. (1973). *Anal. Biochem.* **51**, 604–606.

Robbins, K. C., and Summarai, L. (1973). *Ann. N.Y. Acad. Sci.* **209**, 397–404.

Robinson, H. K., and Leaback, D. H. (1974). *Biochem. J.* **143**, 143–148.

Rodbard, D., and Chrambach, A. (1971). *Anal. Biochem.* **40**, 95–134.

Rodbard, D., Kapadia, G., and Chrambach, A. (1971). *Anal. Biochem.* **40**, 135–157.

Rotbol, L. (1970). *Clin. Chim. Acta* **29**, 101–105.

Salaman, M. R., and Williamson, A. R. (1971). *Biochem. J.* **122**, 93–99.

Shapiro, A. L., Vinuela, E., and Maizel, J. V. (1967). *Biochem. Biophys. Res. Commun.* **28**, 815–820.

Skude, G., and Jeppsson, J. O. (1972). *Scand. J. Clin. Invest. Suppl. 124* **29**, 55–58.

Slater, G. G. (1969). *Anal. Chem.* **41**, 1039–1041.

Smith, I., Lightstone, P. J., and Perry, J. D. (1971). *Clin. Chim. Acta* **35**, 59–66.

Smithies, O. (1959). *Biochem. J.* **71**, 585–587.

Stegemann, H. (1970). *Z. Anal. Chem.* **252**, 165–169.

Ui, N. (1971a). *Biochim. Biophys. Acta* **229**, 567–581.

Ui, N. (1971b). *Biochim. Biophys. Acta* **229**, 582–589.

Ui, N. (1973). *Ann. N.Y. Acad. Sci.* **209**, 198–209.

Vesterberg, O. (1971). *Biochim. Biophys. Acta* **243**, 345–348.

Weber, K., and Kuter, D. J. (1971). *J. Biol. Chem.* **246**, 4504–4509.

Weber, K., and Osborn, M. (1969). *J. Biol. Chem.* **244**, 4406–4412.

Wellner, D., and Hayes, M. B. (1973). *Ann. N.Y. Acad. Sci.* **209**, 34–43.

Williamson, A. R., Salaman, M. R., and Kreth, H. W. (1973). *Ann. N.Y. Acad. Sci.* **209**, 210–224.

Wright, G. L., Farrell, K. B., and Roberts, D. B. (1973). *Biochim. Biophys. Acta* **295**, 396–411.
Wrigley, C. W. (1965). *Aust. J. Biol. Sci.* **18**, 193–195.
Wrigley, C. W. (1968). *Sci. Tools* **15**, 17–23.
Wrigley, C. W. (1970a). *Biochem. Genet.* **4**, 509–516.
Wrigley, C. W. (1970b). *In* "Protides of the Biological Fluids" (H. Peeters, ed.) (*17th Colloq., 1969*), pp. 417–421. Pergamon, Oxford.
Wrigley, C. W., and Shepherd, K. W. (1973). *Ann. N.Y. Acad. Sci.* **209**, 154–162.

6

ISOELECTRIC FOCUSING IN GRANULATED GELS

Bertold J. Radola

Institut für Lebensmitteltechnologie und Analytische Chemie
Technische Universität München
Freising-Weihenstephan, West Germany

I. INTRODUCTION

A. Anticonvective Stabilization of the pH Gradient

After the development of the synthetic carrier ampholytes for isoelectric focusing by Vesterberg (1969) many methodological effects have concentrated on the anticonvective stabilization of the pH gradient. These efforts originated from the notion that stabilization by the vertical columns with density gradients which were employed in some of the first experiments in isoelectric focusing in natural pH gradients does not meet many requirements in both analytical and preparative isoelectric focusing. Long running times and high amounts of analyzed sample and carrier ampholytes are overcome by reducing the size of the column (Weller *et al.*, 1968; Koch and Backx, 1969; Fawcett, 1970; Godson, 1970; Harzer, 1970; Catsimpoolas, 1971a,b; Jonsson *et al.*, 1973; Korant and Lonberg-Holm, 1974). Automatic analysis could eliminate the high expenditure of efforts necessary for evaluation of a single experiment (Jonsson *et al.*, 1969). In situ evaluation of the focusing pattern (Fawcett, 1970; Catsimpoolas, 1971a,b; Fredriksson, 1972; Jonsson *et al.*, 1973) surmounts one of the most severe drawbacks, namely the loss of resolution due to mixing of the separated components on elution (Flatmark and Vesterberg, 1966; Carlström and Vesterberg, 1967; Quast and Vesterberg, 1968), but provides no adequate alternative for determination of the pH gradient and the isoelectric points of the separated components which could be determined only on the eluates with the unavoidable mixing on elution. Use of vertical density gradient columns has been advocated for preparative isoelectric focusing (Vesterberg, 1968, 1971b; Paléus et al., 1971) but here again mixing of the focused components on elution presents an inherent limitation of the technique.

The introduction of gel stabilization must be viewed against the background of these difficulties in work with density gradients. The widespread use of continuously polymerized, homogeneous, or "compact" polyacrylamide gels, as opposed to the granulated gels, for electrophoresis (Gordon, 1969; Maurer, 1971) and the considerable "know-how" in handling these gels undoubtedly contributed to the fact that isoelectric focusing in cylindrical gels (Catsimpoolas, 1968; Dale and Latner, 1968; Fawcett, 1968; Riley and Coleman, 1968; Wrigley, 1968) and in flat bed gels (Awdeh et al., 1968; Leaback and Rutter, 1968) was developed almost simultaneously in different laboratories. Only minor modifications were necessary to adapt polyacrylamide gel electrophoresis to isoelectric focusing with respect to both equipment and visualization techniques. Isoelectric focusing in polyacrylamide gel rapidly proved to be a versatile technique in analytical work, and offered a number of desirable features (cf. Chrambach and Baumann, Chapter 4). However, for proteins with molecular weights greater than 500,000, molecular sieving could present a problem since high gel porosity and satisfactory mechanical strength are two properties of the polyacrylamide gels difficult to reconcile (Finlayson and Chrambach, 1971; Wrigley, 1971; Righetti and Drysdale, 1974). The risk of artifacts caused by polymerization catalysts, described for polyacrylamide electrophoresis (Brewer, 1967; Fantes and Furminger, 1967), pertains also to gel electrofocusing. A prerun is not necessarily a satisfactory way for removal of the polymerization catalysts, and free radicals detectable by electron spin resonance seem to persist even after this treatment (Peterson, 1971). Most applications of isoelectric focusing in polyacrylamide gels are confined to analytical separations. Only few attempts have been made to use cylindrical polyacrylamide gels for preparative fractionations (Finlayson and Chrambach, 1971; Righetti and Drysdale, 1973). Cylindrical gels have an unfavorable geometry for heat transfer which determines the scale of operation and the limits of scale-up. Also only few applications have been described in which flat bed polyacrylamide gels were employed for preparative separation (Leaback and Rutter, 1968; Graesslin et al., 1972; Graesslin and Weise, 1974). A common drawback of preparative isoelectric focusing in polyacrylamide gels is the inefficient recovery of separated proteins, as already well documented by preparative electrophoresis (Gordon, 1969; Rodbard et al., 1974).

B.   Granulated Gels

1.   Systems for Isoelectric Focusing in Granulated Gels

Granulated gels of the Sephadex or Bio-Gel type were introduced for anti-convective stabilization of the pH gradient with the idea of overcoming some of the limitations of both the density gradient technique of isoelectric focusing and stabilization by compact polyacrylamide gels (Radola, 1969, 1971, 1973a,b). While granulated gels have also been used for isoelectric focusing in closed vertical systems with either a cylindrical geometry (Fawcett, 1975) or in flat bed continuous-flow apparatus with cooling from two sides (Fawcett, 1973; Hedenskog, 1975), definite advantages are encountered when working in open horizontal layers. Granulated gels were used first in thin-layer isoelectric focusing for analytical separations. An excellent resolution has been observed for many proteins and enzymes. The separated components were either transferred to a paper print or evaluated by direct densitometry, in both cases without distortion of the focusing pattern and without loss of resolution. The size of the plate and the separation distance could be varied, thus offering together with the available set of carrier ampholytes of different pH ranges a high degree of versatility. Additional advantages of stabilization with granulated gels were found when some specific applications were envisaged, particularly focusing of enzymes followed by detection of enzyme activity.

The flexibility of the dimension of the gel layer permitted the total gel volume to vary within a broad range, and by simply increasing the thickness of the layer it was possible to proceed from thin-layer focusing to preparative separations (Radola, 1971, 1973a, 1975a,b). This property together with the high load capacity of granulated gels extended the range for preparative fractionation from milligram to 10-g amounts of protein without an impairment of resolution. Again, the high versatility in fractionation conditions permitted the method to be optimally adapted to the amount of protein to be separated. Isoelectric focusing in horizontal layers of granulated gels is unique in that it encompasses a wide variety of conditions for the analytical and preparative fractionation of proteins and other amphoteric substances without changing the basic setup.

2.   Properties of Granulated Gels

Granulated gels of different chemical nature are used extensively for gel chromatography, the most frequently employed being the dextran gels of the Sephadex type and polyacrylamide gels of the Bio-Gel series. For both gel materials a large body of information is available on their chromatographic properties (Determann, 1969; Fischer, 1969). In general, most of the properties of these gels, which are desirable when their gel-chromatographic application is considered, also determine their usefulness as an anticonvective medium for isoelectric focusing. Among the properties important for iso-electric focusing are

1. low content of ionic groups,
2. absence of chemical interaction between the gel and the separated substances,
3. chemical stability,
4. well-controlled shape and size characteristics, and
5. printing properties.

Few materials of the rich assortment of supports used in electrophoresis have satisfactory properties for isoelectric focusing. While some of these support-ing media would offer highly desirable operational advantages, e.g., paper, cellulose acetate, or agarose, the majority of them are unsuitable for iso-electric focusing because of an excessive content of charged groups.

a.   Chemical Properties.   Both the Sephadex and Bio-Gel types of gran-ulated gels contain only low levels of charged groups, mainly carboxylic groups (Determann, 1969; Fischer, 1969). The carboxyl group content of $20-30$ μequiv/g of dry gel found for some of the Sephadex gels in early appli-cations (Flodin, 1962) has been subsequently much reduced. Various G types of the Sephadex gels do not differ in their carboxyl group content referred to the dry gel. According to a laboratory manual issued by Bio-Rad Labora-tories (1971) the approximate total cation exchange capacity of the Bio-Gel P products is less than 0.05 μequiv/g of dry material regardless of the cross-linking. Cationic groups are claimed not to exist in the Bio-Gel material. The low content of ionic groups in the Sephadex and Bio-Gel types of granulated gels has already facilitated their application as an excellent anticonvective medium in zone electrophoresis (Fischer, 1969). Parenthet-ically, the Sephadex gels were originally developed for use in column electrophoresis. The electroosmotic flow is very low with the granulated gels and heat dissipation is better than with other supporting media since the gel particles and not only the interstitial liquid conduct the current. These are also useful properties of the granulated gels when applied for anticon-vective stabilization in isoelectric focusing.

Other granulated gels employed in gel chromatography, e.g., granulated agarose (Hjertén, 1964) and porous glass (Haller, 1965), are unsuitable for isoelectric focusing owing to the high content of charged groups. Homo-geneous agarose gels were tried in some early experiments in isoelectric focusing (Riley and Coleman, 1968; Catsimpoolas, 1969) but have found only very limited use despite other desirable properties (Quast, 1971), espe-cially in combination with immunological detection methods. Chromato-graphically purified agarose is better suited for isoelectric focusing than hitherto available commercial preparations of agarose (Johansson and Hjertén, 1974) but at present it is not clear whether constant isoelectric point (pI) values can be obtained with these gels, nor is this material available in a granulated form.

Interaction between the gel support and the separated substances could lead either to irreversible binding or in case of reversible interactions to erroneous pI values since the attainment of equilibrium conditions would be retarded to an undefined degree. Fortunately, only few examples have been known in gel chromatography for which the separation is not a function of the molecular dimension of the separated material. Thus, irreversible interactions can be expected to be absent in isoelectric focusing also. In work with substances containing, e.g., aromatic structures for which reversible adsorption has been described in gel chromatography, possible interactions should be taken into account in isoelectric focusing also.

Chemical stability of the granulated gels is good and degradation of the gels is unlikely to occur at the most frequently employed pH ranges in isoelectric focusing in horizontal layers. The Sephadex gels are stable even on prolonged exposure to 0.02 M HCl or 0.25 M NaOH; the Bio-Gel products should not be exposed to pH levels below 2 or above 10. Extreme pH values may be necessary in some special applications or may be encountered in restricted regions close to the electrodes. When compared with the rather long exposure of the gel packings during column operation in gel chromatography, contact with extreme pH values is unlikely to have a detrimental effect on the gels during isoelectric focusing. Different conditions in this respect are found in the continuous-flow system in the neighborhood of the electrodes (Fawcett, 1973).

b. Physical Properties.   The shape and size characteristics of granulated gels are known to influence resolution and flow rates strongly in gel chromatography, and are also of paramount importance in isoelectric focusing. Spherical particles of small diameter and narrow particle size distribution are essential in isoelectric focusing for achieving high resolution. Best results are obtained with gel particles with a diameter <40 μm which for the Sephadex gels are available as the superfine grade and for the Bio-Gel series as the "-400" mesh product. Resolution and zone sharpness could probably be improved by using materials with smaller particle diameter and a narrower particle size distribution. At least in the technique employing horizontal gel layers the flow characteristics of the granulated gels are of minor importance.

A great assortment of granulated gels both of the Sephadex and Bio-Gel series is available for different fractionation ranges. In some applications it could also prove useful in isoelectric focusing to take advantage of the different exclusion limits of these gels. In Table I the fractionation ranges of the gels together with some technical data are given. The data on bed volume

TABLE I

Technical Data for Sephadex and Bio-Gel Gels

Product	Particle diameter (μm)	Fractionation range	Volume of gel suspension (ml/g Xerogel)	Printing properties
Sephadex				
G-25 superfine	10–40	1,000–5,000	5	Bad
G-50 superfine	10–40	1,500–30,000	10	Good
G-75 superfine	10–40	3,000–70,000	14	Excellent
G-100 superfine	10–40	4,000–150,000	16	Good
G-150 superfine	10–40	5,000–400,000	20	Satisfactory
G-200 superfine	10–40	5,000–800,000	25	Unsatisfactory
Bio-Gel				
P-4 – 400 mesh	< 37	800–4,000	4.5	Good
P-6 – 400 mesh	< 37	1,000–6,000	5.5	Good
P-10 – 400 mesh	< 37	1,500–20,000	10.5	Good
P-30 – 400 mesh	< 37	2,500–40,000	11.5	Good
P-60 – 400 mesh	< 37	3,000–60,000	14.5	Excellent
P-100 – 400 mesh	< 37	5,000–100,000	15.5	Good
P-150 – 400 mesh	< 37	15,000–150,000	17	Satisfactory
P-200 – 400 mesh	< 37	30,000–200,000	25	Satisfactory

provide a rough idea of the volume of gel suspension which can be prepared for either coating the thin-layer plates or for preparative experiments.

Mechanical rigidity of the granulated gels appears relatively unimportant, at least when granulated gels are used in horizontal layers. With the exception of the preparation of gel suspension the granulated gels are not subjected to fracture in isoelectric focusing during normal operating conditions. No liquid flow is established in the horizontal systems, but in the continuous-flow systems (Fawcett, Chapter 7) gels with high exclusion limits could create difficulties comparable to those known from column gel chromatography.

The printing properties are important for evaluation by taking a print from the gel layer, a step of the method to be described in more detail in Section II.A.4. Granulated gels differ strongly with respect to printing properties. Some gels are unsuitable for printing, e.g., Sephadex G-25, because insufficient interstitial liquid is transferred to the paper print, others tend to cause trouble because variable amounts of gel adhere to the paper. Surprisingly, gels with suitable printing properties in thin-layer gel chromatography may prove unsatisfactory when employed after isoelectric focusing.

c. Molecular Sieving. In contrast to some electrophoretic methods in which molecular sieving is a desirable property which contributes to their high resolving power (Gordon, 1969; Maurer, 1971), nonsieving gels are preferable in isoelectric focusing which is an equilibrium method and in which all amphoteric species should reach their p*I* before the pH gradient decays. Whereas in compact polyacrylamide gels the proteins will be retarded more strongly with an increase in molecular dimensions, in granulated gels all molecules above the exclusion limit will pass the gel without sieving. Thus, these gels appear particularly suitable for isoelectric focusing of substances with molecular weights $> 500,000$ for which it could prove difficult to reach equilibrium conditions in compact polyacrylamide gels (Wrigley, 1971; Righetti and Drysdale, 1974). Small molecules will be retarded on isoelectric focusing in granulated gels, depending both on size and shape of the separated molecules, and on the exclusion limit of the gel. In retardation depending on molecular sieving a basic difference exists between the granulated and the compact gels. Whereas in the granulated gels retardation is completely "reversible," in the compact gels the larger molecular weight components may be retarded irreversibly by clogging the limiting pores of the gels.

d. Advantages of Granulated Gels. In view of the chemical and physical properties granulated gels of the Sephadex or Bio-Gel type appear well suited to function as an anticonvective medium in isoelectric focusing. Among the performance properties the more important advantages of these gels may be summarized as follows:

1. excellent resolution,
2. high loading capacity,
3. evaluation by printing,
4. great flexibility with respect to the dimension of the gel layer,
5. use for analytical and preparative separations in the same system, and
6. simple and quantitative elution of the separated substances from the gel.

These properties are treated in more detail in the following sections.

Granulated gels afford the great advantage of being available ready for use without need for polymerization. The risk of artifacts caused by polymerization catalysts, discussed for the compact polyacrylamide gels, is thus avoided. Unreacted materials are essentially nonexistent in the commercially supplied material. Granulated gels are a very convenient material because no polymerization vessels are needed by which the dimension of the gels is determined.

The handling of granulated gels is simple and a certain familiarity with this material can be assumed as a result of the widespread use of gel chromatography. Some experience in work with thin-layer gel chromatography (Johansson, 1971) would provide a good background for isoelectric focusing in granulated gels. Even without previous experience, work with granulated gels is simple and does not require special skill.

II. THIN-LAYER ISOELECTRIC FOCUSING

When gels are employed for anticonvective stabilization of the pH gradient many arguments are in favor of using layers instead of gels with a cylindrical geometry:

1.   Many samples can be analyzed simultaneously under identical conditions, thus facilitating their comparison.

2.   The preparation and handling of gel layers is simpler than that of cylindrical gels.

3.   The gel layer is accessible at any site for sample application and pH measurements during the experiment and after completion of the run.

4.   An efficient cooling can be maintained in a layer, which is important with respect to the applicable voltage, the attainable resolution, as well as in view of the possible detrimental effect of extreme pH values on some proteins.

In contrast to most electrophoretic methods in which the separation distance is determined by the dimensions of the apparatus a high degree of flexibility is offered by thin-layer isoelectric focusing. In its simplest form the setup for thin-layer isoelectric focusing consists of a gel layer on a cooled glass plate on which two electrodes are placed on both ends through which the voltage is applied. A setup of this type can be easily constructed and is also available as a commercial apparatus with adjustable electrodes.* Thus the interval between the electrodes can be varied to obtain any desirable separation distance which can be optimally adapted to the requirements of the separation task. With granulated gels an added advantage is provided by the circumstance that the size of the plate is not determined by the dimension of the polymerization cuvettes which are needed for the preparation of the flat polyacrylamide gels.

A.   Isoelectric Focusing of Proteins

1.   Comparison of Gel Stabilization with a Density Gradient

Optimum separation parameters for thin-layer isoelectric focusing are most easily determined using mixtures of defined pH marker proteins with isoelectric points distributed over a suitable pH range (Radola, 1973a,b). A mixture containing the following eight proteins has been found particularly suitable: ferritin, bovine serum albumin, β-lactoglobulin, conalbumin, horse and sperm whale myoglobin, ribonuclease, and cytochrome c. The colored proteins of this mixture provide a very convenient control during focusing in the acidic (ferritin), neutral (myoglobin), and alkaline (cytochrome

---

* Desaga, Heidelberg, West Germany.

c) pH range. Single proteins and mixtures of a few of these marker proteins are useful for testing narrow pH ranges of the carrier ampholytes. The pattern shown in Fig. 1 clearly demonstrates the excellent resolution obtained in thin-layer isoelectric focusing. Cytochrome *c*, ribonuclease, both myoglobins, and conalbumin show a characteristic and well-reproducible

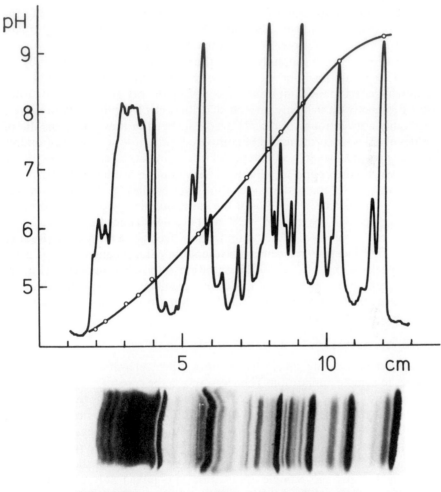

FER BSA LAC   CON   MYH MYW RIB CYT

Fig. 1. Thin-layer isoelectric focusing of pH marker proteins in pH 3–10 carrier ampholytes. Sephadex G-75 superfine; thickness of the gel layer, 0.06 cm; separation distance, 20 cm. Bottom: paper print stained with Coomassie Brilliant Blue G-250. —, densitometric tracing; ○, pH gradient. Marker proteins: FER, ferritin; BSA, bovine serum albumin; LAC, β-lactoglobulin; CON, conalbumin, MYH, horse myoglobin; MYW, sperm whale myoglobin, RIB, ribonuclease; CYT, cytochrome *c*.

banding into a major and several minor components. The remaining proteins also consist of several components. The pattern of a protein does not depend on whether it is run as a single protein or in a mixture. It may be concluded from this finding that a uniform pH gradient is formed in the gel layer, providing the basis for comparison of different proteins separated on a single plate.

Since proteins are ampholytes it can be expected that they modify the course of the pH gradient. This effect becomes visible only at high loads when the amount of protein contributes a substantial fraction to concentration of the carrier ampholytes on their migration track. Generally, at protein loads optimal for staining with most currently used dyes, this modifying effect of the sample on the course of the pH gradient can be ruled out. Proteins with a highly uneven distribution of components are likely to modify a given range of the pH gradient. Even in these instances the pI values of unknown proteins can be compared with those of the corresponding marker proteins owing to a tendency of line fusing at the ends of the colored zones which correspond to regions of very similar or identical pH.

Comparison of the pattern after thin-layer isoelectric focusing with that obtained on isoelectric focusing of the same mixture (with ferritin omitted because of its high absorbance at 280 nm) in a density gradient is instructive because the advantages of gel stabilization become very evident (Radola, 1973b). After focusing in a density gradient the pattern consists of just a few major peaks with only poor differentiation into subcomponents (Fig. 2). Loss of resolution on focusing in density gradients results from mixing of the focused proteins on elution. For the colored proteins a good resolution is observed in the column in situ.

2. Sample Application

Isoelectric focusing differs from most other separation methods with respect to sample application. The substance to be separated can be applied as a very dilute solution, since it becomes concentrated during the experiment as a result of the focusing effect. While in thin-layer isoelectric focusing the sample can also be applied by mixing it with the gel suspension used for coating the plate, it is preferable to apply the sample on the preformed gel layer. Open horizontal gel layers afford the advantage that the sample can be applied at any site. Sample application in compact polyacrylamide gels may present some difficulties because with the most convenient way of application (by means of a paper strip), variable amounts of proteins have been found to be retained by adsorption (Wadström and Smyth, 1975). In work with granulated gels the sample is applied as a solution directly on the gel layer which easily soaks up the sample up to 50 μl/cm of an analytical 0.06-cm layer. Greater volumes can be applied successively in several portions in the direction of the migration track. A low salt content of the sample

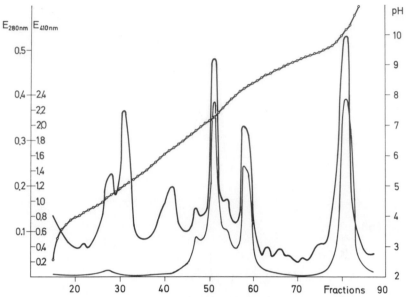

Fig. 2.   Isoelectric focusing in a sucrose density gradient column in pH 3–10 ampholytes. Proteins: human serum albumin, β-lactoglobulin, conalbumin, horse myoglobin, sperm whale myoglobin, ribonuclease, and cytochrome c, 5 mg of each. Focusing at 300 V for 48 hr. ○, pH gradient; —, absorbance at 280 nm; —, absorbance at 410 nm. [From Radola (1973b).]

(<0.005 M) is essential for getting satisfactory focusing patterns. In work with crude material, viscosity of the sample that is too high often results in distorted patterns. While in some cases it is the substance to be analyzed that causes the excessive viscosity, most frequently it is due to contaminating polysaccharides. Since the removal of these polysaccharides by ion exchange or gel or affinity chromatography could prove difficult, just simple dilution should be tried, which in many instances may suffice to eliminate the undesirable viscosity effect.

The site of application is optional but as a rule application of the sample at a certain distance from the final position of the major components will cause them to migrate into a gradient and sharpen the zones. Depending on its pH stability, the sample should not be applied too close to the electrodes to avoid denaturation (Lewin, 1970). For components focused at extreme pH values the time of exposure of these components to a possibly detrimental pH after reaching equilibrium should be reduced. A way of achieving this is to prefocus the system and to apply the sample after the pH gradient has been already established. The risk of artifacts caused by exposure to low pH is usually less than in the density gradient technique, as a result of the shorter focusing time and efficient cooling, but should be considered carefully, especially in work with enzymes.

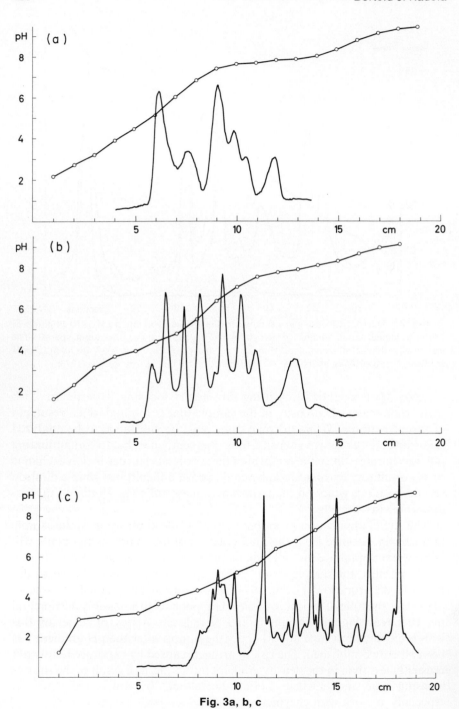

Fig. 3a, b, c

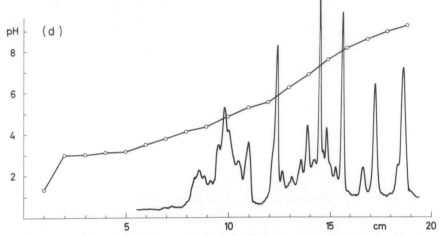

Fig. 3.   The effect of time and voltage on the protein pattern and pH gradient in thin-layer isoelectric focusing. (a) At 200 V, 2 hr; (b) 200 V, 4 hr; (c) 200 V, 4 hr and 800 V, 2 hr; (d) 200 V, 4 hr and 800 V, 4 hr. Plate 20 × 10 cm, pH 3–10 ampholytes. Proteins (major peaks) from left to right: ferritin, human serum albumin, β-lactoglobulin, conalbumin, horse myoglobin, sperm whale myoglobin, ribonuclease, and cytochrome *c*. Mixture containing 1% of each protein applied in the middle of the plate. Staining of the print with Light Green SF; —, densitogram; ○, pH gradient. [From Radola (1973b).]

3.   Size of the Plate, Voltage, and Time

Flexibility is a feature in thin-layer isoelectric focusing in granulated gels. While standard thin-layer plates (20 × 20 or 20 × 10 cm) with a separation distance of 20 cm are convenient for most experiments, shorter separation distances, e.g., 10 cm or longer up to 40 cm, have proved useful for some special applications (Radola, 1973b). With the adjustable electrodes any intermediate separation distance may also be chosen. For proteins with close isoelectric points the highest resolution is obtained when narrow pH range ampholytes are used on a 40-cm separation distance. The voltage and time necessary for obtaining optimum results depend on the separation distance, the pH range of the carrier ampholytes, and to some degree also on the properties of the analyzed substance. Comparatively low-voltage gradients of 7–10 V/cm in the first phase of focusing in pH 3–10 carrier ampholytes are essential for a good quality pattern. Voltage gradients of 30–50 V/cm may be used in the final phase of focusing when most proteins have reached positions close to their p*I* values. The effect of voltage and time on the pH gradient and the protein pattern is shown in Fig. 3. Voltage gradients of > 100 V/cm have been advocated for flat bed isoelectric focusing in compact polyacrylamide gels (Söderhalm *et al.*, 1972). While it is possible to work also in granulated gels at these high-voltage gradients their usefulness remains questionable since they generally impair the quality of the focusing pattern.

The effect of voltage and time on the pH gradient must be considered carefully when pH measurements after isoelectric focusing are planned. Constant isoelectric points have been observed for pH marker proteins under strongly varying experimental conditions of time, voltage, and site of application (Table II). Proteins with isoelectric points of < pH 8.5 reach the same position on a 20-cm plate after 6–8 hr of focusing, irrespective of their application site.

TABLE II

Isoelectric Points of pH Marker Proteins[a]

| Protein | \multicolumn{5}{c}{Thin-layer isoelectric focusing} | Focusing in a sucrose density gradient |
	200 V, 4 hr; 800 V, 2 hr	200 V, 4 hr; 800 V, 4 hr	200 V, 4 hr; 1000 V, 4 hr	200 V, 16 hr; 800 V, 2 hr	200 V, 16 hr; 800 V, 4 hr	
Cytochrome c	9.30	9.28	9.29	9.25	9.30	10.16[b]
Ribonuclease[c]	8.85	8.91	8.89	8.85	8.92	—
Myoglobin[d]						
sperm whale						
Major component	8.14	8.17	8.21	8.18	8.19	8.19[e]
Minor component	7.66	7.67	7.67	7.69	7.70	7.65
Myoglobin, horse[d]						
Major component	7.31	7.33	7.34	7.32	7.34	7.35[e]
Minor component	6.83	6.85	6.84	6.88	6.89	6.89
Conalbumin[c]	5.86	5.88	5.89	5.88	5.89	6.03[b]
β-Lactoglobulin[f]	5.13	5.14	5.15	5.14	5.14	—
Bovine serum[g]						
albumin I	4.86	4.85	4.85	4.84	4.85	5.07[b]
II	4.74	4.72	4.76	4.71	4.70	—
Ferritin I	4.51	4.48	4.52	4.50	4.50	4.50
II	4.42	4.37	4.38	4.37	4.38	—
III	4.23	4.25	4.27	4.22	4.20	—

[a] Determined by thin-layer isoelectric focusing in pH 3–10 ampholytes, focusing on 20 × 10 cm plates in a 0.06-cm layer of Sephadex. Fifty microliters of the mixture of eight proteins, each at 1%, was applied in the middle of the plate. The pH measurements were at 25°C on 20–30 μl gel samples removed as 20 × 2 mm streaks from the layer and liquefied with an equal volume of distilled water. Average values from four separate runs. The standard deviation was 0.01–0.06 (Radola, 1973b).

[b] Determined by focusing of a mixture of proteins in pH 3–10 ampholytes. The albumin and β-lactoglobulin were partially overlapping. The ferritin was not resolved into subcomponents.

[c] Visible in the gel after focusing as a transparent zone.

[d] The pI values were determined for the two quantitatively predominant components.

[e] Focusing in pH 7–9 (sperm whale myoglobin) and pH 6–8 (horse myoglobin) ampholytes. Horse myoglobin II emerged from the column as a single peak.

[f] Consists of two main components. The pI was measured only for the isoelectrically precipitated component; the pI of the other was 0.15–0.2 higher.

[g] Bromphenol Blue added; the pI values were measured for the stained zones.

4. The Print Technique

When working with the granulated gels, proteins are most conveniently detected by the print or replica technique (Radola, 1968; Johansson, 1971). After focusing, a sheet of a dry chromatographic paper is rolled onto the gel layer. On short contact, the paper soaks up the liquid phase with the separated proteins; it is then removed from the gel layer, dried, and stained. When colored proteins are separated it can be seen that the focused proteins enter from the gel layer into the print essentially without distortion (Radola, 1973b). The print technique was previously used for detection of proteins and enzymes in work with other anticonvective supports such as starch grains, polyvinyl chloride, copolymers of polyvinyl chloride and acetate, and also with some compact gels such as agar or polyacrylamide gels (Bloemendal, 1963). However, optimal results are obtained when the print technique is applied to the Sephadex or Bio-Gel type of granulated gels of sufficiently small particle size. The paper print is prepared with chromatographic papers of suitable surface weight, mechanical strength, and destaining properties. In thin-layer isoelectric focusing papers with approximately $120–180$ g/m^2 give best results; with thicker gel layers and correspondingly higher content of the liquid phase, thick papers with up to 350 g/m^2 should be used. For some special visualization reactions glass-fiber papers may also be used. Without printing, proteins can be visualized in the gel layer with 8-anilino-1-naphthalenesulfonic acid (Merz *et al.*, 1974a). A dilute water solution of the reagent is sprayed on the gel; after a few seconds the proteins are visible as fluorescent bands.

a. Staining. Most of the dyes employed for staining in zone electrophoresis can also be adapted for the print technique after isoelectric focusing. Of a number of dyes tested, Coomassie Brilliant Blue G-250 proved to be the best. This dye combines a high sensitivity and a perfectly destained background with a reasonable destaining time. From 0.25 to 0.5 μg of protein is detected per centimeter of an isoelectrically focused zone, a sensitivity which compares well with that described for other applications (Fazekas de St. Groth *et al.*, 1963; Chrambach *et al.*, 1967; Bramhall *et al.*, 1969). At equal sensitivity Coomassie Brilliant Blue G-250 gives a more uniform staining than Coomassie Brilliant Blue R-250 which appears to be preferred in most staining procedures for compact polyacrylamide gels (Graesslin *et al.*, 1971; Vesterberg, 1971b; Söderholm *et al.*, 1972). Experiments with different dyes, Amido Black 10 B, Bromphenol Blue, Coomassie Brilliant Blue G-250 and R-250, Coomassie Violet R-150, and Light Green SF, at concentrations of 0.1, 0.2, 0.5, and 1%, have shown that all dyes form insoluble or poorly soluble complexes with the Ampholine carrier ampholytes at higher concentrations (Radola, 1973b). While unsuitable for protein staining a 0.5–1% concentration of some dyes, particularly of Light Green

SF, proved useful for visualization of the carrier ampholytes. A total of about 50–65 bands could be stained on a 40-cm plate; these bands focused sharply in the acidic pH range and were more diffuse in appearance at the neutral and alkaline pH. These results are in good agreement with other detection methods for Ampholine patterns, e.g., by glucose caramelization (Felgenhauer and Pak, 1973).

Since the carrier ampholytes interfere with staining they must be removed by washing with trichloroacetic acid, sulfosalicylic acid, or mixtures of both. With some batches of the Ampholine carrier ampholytes difficulties are encountered in destaining the region between pH 6 and 8, a broad zone of incompletely destained background being observed even on prolonged destaining. Experiments with a new type of carrier ampholytes, Servalytes,* indicate that they form insoluble or poorly soluble complexes with different dyes to a much lesser degree (Radola, unpublished results). Use of these carrier ampholytes facilitates staining of proteins after isoelectric focusing, as judged by much shorter destaining time and no residual background staining. As a rule destaining times of paper prints are reasonably short when compared with destaining of compact polyacrylamide gels. Excessively long destaining times because of incomplete removal of carrier ampholytes and formation of insoluble complexes with the dye have been experienced initially in work with compact polyacrylamide gels (Righetti and Drysdale, 1974). Modified staining procedures are now available for staining compact polyacrylamide gels, but even these techniques require more time than staining of the paper prints.

In addition to protein staining, other detection methods, e.g., for glycoproteins and/or lipoproteins, can be carried out on the print (Delincée and Radola, 1971). Parts of a broad print can be stained by different methods and conveniently compared after staining. Soluble carbohydrates which after focusing in Sephadex gels are transferred with the interstitial liquid to the paper are easily removed simultaneously with the carrier ampholyte by washing with trichloroacetic acid. Alternatively, Bio-Gel may be employed when detection of glycoproteins is considered.

The stained paper print is a very convenient document which can be preserved easily and kept for later inspection. The advantage is obvious, particularly in view of the storage problem engendered by an increasing number of cylindrical or flat bed gels, which are also plagued by bleaching of the stained zones with time. The stained paper print can be evaluated densitometrically by most commercially available densitometers, e.g., those suitable for evaluation of paper electropherograms or thin-layer chromatograms. The densitometry is only semiquantitative for two reasons. Deviations

---

* Serva, Heidelberg, West Germany.

from the Lambert–Beer law are to be expected for some of the highly concentrated zones. The separated proteins are not transferred quantitatively from the gel layer into the paper, because depending on the exclusion limit of the gel variable amounts of nonexcluded proteins will be retained by the gel (Radola, 1968). Location of radioactivity in the paper print by a strip scanner has been reported (MacGillivray and Rickwood, 1975).

b.   Direct Densitometry.   An alternative to detection of proteins by the print technique is direct densitometry in the gel layer, either on ordinary glass plates (Radola and Delincée, 1971) or on quartz plates (Radola, 1975a). This approach requires much higher concentrations of proteins than those necessary for staining. In some applications direct ultraviolet densitometry could prove useful, particularly in work with single proteins for which variations of specific extinction coefficients are absent. If the chromophoric groups are suitable, direct densitometry could be the method of choice in work with peptides, which are difficult to visualize after focusing because of interference from the carrier ampholytes.

5.   Comparison of Different Gels

Sephadex and Bio-Gel gels with different exclusion limits and particle sizes were tested (Radola, 1973b). The Sephadex superfine gels with a dry particle diameter of 10–40 μm, and the Bio-Gel product -400 mesh corresponding to a particle diameter <40 μ gave decidedly better separations than the gel grades with coarser particle sizes. Sephadex G-25 and to some degree also G-50 gave irregular patterns, particularly in work with the myoglobins and cytochrome *c*. These irregularities could result from an interaction of the tightly cross-linked gel with the proteins, an effect comparable to what is known for some peptides containing aromatic residues (Determann, 1969; Fischer, 1969). It is a question whether in isoelectric focusing this disturbance can be attenuated, e.g., by high concentrations of urea. In addition to these irregularities Sephadex G-25 presents difficulties on printing because of low content of interstitial liquid. Sephadex G-75 superfine was found to be the most suitable supporting gel. Even better separations than those obtained with G-75 are achieved with Sephadex G-100, G-150, and G-200 owing to an increased spreading, which improves with increasing G numbers. A satisfactory print cannot always be obtained with these gels because variable amounts of the gel adhere to the paper print and subsequently complicate its drying and staining.

The granulated polyacrylamide gels of the Bio-Gel series are generally handled less easily than the Sephadex gels. The Bio-Gel materials tend to clot with the result that protruding particles are frequently visible in the gel layer. More homogeneous suspensions can be obtained using high-speed homogenizers, although at the expense of some fracture of the gel particles which may adversely affect resolution. The low-numbered Bio-Gels, P-4, P-6, and P-10, could well replace the Sephadex G-25 and G-50 gels if there is a need for gels with a fractionation range of 500–17,000. Bio-Gels could be of interest in those applications in which either an interaction of the separated substance with the gel matrix or contamination with soluble carbohydrates is to be avoided. The higher G-numbered Sephadex gels may contain up to 10% of free dextran which could interfere with some detection methods or contaminate the eluate in preparative runs. Also in work with some enzymes, e.g., glycoside hydrolases, the Bio-Gels could be preferable.

Granulated agarose gels have not yet been used for thin-layer isoelectric focusing. Their use depends primarily on a sufficiently strong reduction of the electroendosmotic effect owing to residual charged groups.

6. Isoelectric Precipitation

In layers of granulated gels the proteins precipitated at the isoelectric point are immobilized in the gel. Even relatively high amounts of iso-electrically precipitated proteins do not affect the focusing pattern of the accompanying proteins (Radola, 1973b). When increasing amounts of β-lactoglobulin, a protein that precipitates at its isoelectric point, are applied on a plate together with other pH marker proteins white zones of precipitated β-lactoglobulin can be clearly seen in the gel layer. The zones of precipitated proteins are usually very regular in appearance and can be easily recognized in the pattern. Frequently the precipitated protein is in equilibrium with the soluble protein which can be detected by the print technique.

7. Load Capacity

An outstanding property of thin-layer isoelectric focusing is the high load capacity which permits separation of proteins in amounts considerably exceeding those handled by conventional electrophoretic methods. Also, for isoelectric focusing in cylindrical polyacrylamide gels a much higher load capacity has been found than in disk electrophoresis (Finlayson and Chrambach, 1971). In an analytical gel layer (thickness ~ 0.06 cm) regular protein zones are obtained up to 1 mg of a single protein per centimeter of focused zone, a value which is higher than that optimal for staining by a factor of 10–30. Overloading results first in irregular zones which oc-casionally disintegrate with droplet formation. The weakly acidic and near neutral ranges between pH 4 and 5 appear to be more sensitive to overloading than the remaining part of the pH gradient. High protein loads are important when less sensitive detection methods, e.g., for glycoproteins and lipo-proteins, are considered. They could prove useful also in those applications in which the components of interest occur only in small quantities and are accompanied by a high excess of contaminating components. Taking into account the load capacity of thin-layer isoelectric focusing and the sensitivity of protein detection by staining it should be possible to detect as little as 0.01–0.05% of a minor component in the presence of an excess of other proteins, provided that the components of interest are sufficiently spaced in the pH gradient, a requirement which can be at least partially approached by choice of an optimal pH range of the carrier ampholytes and also through variation of the separation distance.

8.   Principle of Localized Spreading

With many protein systems of biological origin isoelectric patterns are observed with pI values of the separated components spread over a wide pH range with local accumulation of a great number of components in a rather narrow pH range. In some applications it would obviously be an advantage if the crowded part of the pattern could be resolved more efficiently while simultaneously preserving the remaining part of the pattern. To achieve this the principle of localized spreading can be applied by mixing the pH 3–10 carrier ampholytes with narrow pH range ampholytes at different ratios (Radola, 1973b). A pH gradient can thus be achieved consisting of a section with a flat course corresponding to the narrow range ampholytes and steeper section(s) in the remaining part. This principle of localized spreading adds to versatility of isoelectric focusing by providing a basis for modifying the separation conditions in conformity with the particular properties of the analyzed material.

9.   Plateau Phenomenon

On prolonged focusing in granulated gels a cathodic drift is observed which results in a progressive flattening of the pH gradient in the region of neutrality. This flattening of the pH gradient, termed the plateau phenomenon (Finlayson and Chrambach, 1971), has been described for all systems of isoelectric focusing (Haglund, 1975) and has been studied in some detail with the compact polyacrylamide gels (Bates and Deyoe, 1972; Miles et al., 1972). The mechanism of the plateau phenomenon is not adequately explained. In granulated gels it is at least partially due to the residual content of charged groups. Treatment of Sephadex gels with propylene oxide (Aspinall and Cañas-Rodriguez, 1958; Zitko and Bishop, 1966), which is known to reduce the content of carboxyl groups, noticeably diminished the electroendosmotic shift also (Radola, 1973b). While present in granulated gels the cathodic drift does not exclude equilibrium conditions as judged by constant pI values of different proteins under varying experimental conditions (Table II). Under these conditions the relatively uniform, nominal pH gradients cover 70–90% of the total length of the gel layer. Increasing the concentration of the carrier ampholytes, which in granulated gels are used at a 1% (w/v) concentration, diminishes the cathodic drift. At higher concentrations of the carrier ampholytes a greater portion of the separation distance will be covered by the nominal pH gradient. Arginine and lysine, which were also present in older preparations of Ampholine carrier ampholytes, were added to the gel to reinforce the alkaline end of the pH gradient (Radola, 1973b). Addition of these basic amino acids is not necessary for the currently available carrier ampholytes, particularly not for the Servalytes which have an extended range reaching pH 11.

The cathodic drift may under unfavorable conditions lead to a loss of components with extreme by alkaline pI values. Whereas in gel systems for isoelectric focusing with the pH 3–10 Ampholine carrier ampholytes pH values of 9.3–9.5 are obtained at the alkaline end of the pH gradient, with the pH 2–11 Servalyte carrier ampholytes pH values near 10.5–10.7 are measured in the gel layer, close to the cathode. In work with substances isoelectric at extremely basic pH values improved conditions are provided by

(a)   focusing in an inert gas atmosphere, e.g., argon or nitrogen,

(b)   use of concentrated electrode solutions at the cathode, e.g., 2–4 M ethylenediamine or sodium hydroxide,

(c)   placing the electrodes in the extension of the gel layer and not on the top of its end, and

(d)   addition of pH 9–11 carrier ampholytes to the broad range ampholytes to reinforce their concentration at the akaline end of the pH gradient.

B.   Isoelectric Focusing of Enzymes

The separation and characterization of enzymes represent one of the most important applications of isoelectric focusing. Despite the availability of gel techniques a very substantial part of isoelectric focusing of enzymes is still carried out in systems stabilized by density gradients. In addition to the disadvantages of anticonvective stabilization by density gradients pertinent to isoelectric focusing in general (Section I.A), in work with enzymes, the determination of activity in the collected fractions, usually amounting to 50–200 per single column, imposes a severe limitation which certainly excludes the method from many potential applications. A further drawback is the interference of sucrose in the assay of a variety of soluble and particulate enzymes (Hinton et al., 1969).

Gel media are gaining increasing importance for isoelectric focusing because they offer an unsurpassed resolution and simple detection of activity by staining in situ (Wadström and Smyth, 1975). Isoelectric focusing followed by visualization of the enzyme activity has been described in some of the first applications of the method (Dale and Latner, 1968; Leaback and Rutter, 1968; Delincée and Radola, 1970a). On isoelectric focusing of enzymes in gel media, visualization is generally preferred to activity determinations in eluates from gel segments. Enzyme visualization renders isoelectric focusing in gels also attractive for routine work because of a considerable reduction of efforts in evaluation of the experiments.

Enzyme visualization methods can be divided into several categories according to the nature of the reagent used:

1.   chromogenic, in which a colorless substrate is converted by the enzyme into a colored product;

2.   fluorogenic, which depends on the enzymatic generation of a fluorescent substrate or vice versa;

3.   chemical, which uses chemical reagents to couple with the products of the enzyme reaction to form highly colored compounds or to stain the undegraded substrate;

4.   electron transfer dyes (e.g., dianisidine, Thiazolyl Blue-MTT), which participate in enzyme-catalyzed oxidation–reduction reactions to form colored products;

5.   enzyme-linked stains, which use exogenous enzymes in staining mixtures to trap and change the product of the enzyme reaction under analysis into a product which can be detected colorimetrically, fluorometrically, etc., by any of the methods just mentioned;

6.   radioactive, which detects the products of the enzyme reaction by autoradiography.

Most visualization reactions described for enzyme staining following enzyme separation by electrophoretic methods (Latner and Skillen, 1968; Clausen, 1969; Maurer, 1971) can be easily adapted to isoelectric focusing. A pH adjustment is usually necessary since the pI values of the separated enzymes will as a rule not coincide with their pH optimum of activity. Enzyme visualization reactions have been developed mainly for compact gels such as agarose, starch, and polyacrylamide gels. Reactions requiring low-molecular-weight substrates and cofactors are most simply carried out by incubating the compact gel after separation of the enzymes either in a single solution or sequentially in several solutions containing the reagents for visualization. Detention of enzymes acting on macromolecular substrates is less simple, four approaches being available:

1.   incorporation of substrate into the gel prior to the separation, a method of limited applicability because of possible interference with the separation process;

2.   diffusion of the substrate into the gel after separation, a method which suffers from the drawback that the simultaneous diffusion of the separated

enzymes will result in a loss of resolution, e.g., for components insufficiently spread in the gel or for minor components adjacent to major ones (Vesterberg, 1973);

3. overlayering with another gel containing the substrate in a preformed layer (Vesterberg and Eriksson, 1972);

4. overlayering with an agarose solution which solidifies on the surface of the running gel, thus ensuring better transfer of the enzymes into the substrate gel layer than by the preceding method (Arvidson and Wadström, 1973; Wadström and Smyth, 1975).

1. Detection of Enzyme Activity by the Print Technique

After thin-layer isoelectric focusing in granulated gels detection of enzyme activity is accomplished by the print technique (Delincée and Radola, 1970a; Radola, 1973a). Enzyme detection is less complicated than protein staining because the carrier ampholytes do not interfere with most visualization reactions and therefore must not be removed. Detection of enzyme activity by the print technique consists of the following steps:

1. A buffer of sufficient buffering capacity and corresponding to the optimal pH of the enzyme is incorporated into the paper for the print.

2. The buffered paper is then impregnated with the substrate preferentially dissolved in an organic solvent. Water-soluble substrates of sufficient stability can be incorporated into the paper simultaneously with the buffer.

3. The dry buffered and substrate-impregnated paper is placed on the gel layer. During short contact of a few seconds up to a few minutes the paper soaks off the liquid phase with the separated enzymes of the gel layer.

4. The enzyme reaction proceeds in the paper print which is removed from the gel layer and dried either at room or an elevated temperature. The enzymes are visualized in this step or on applying an additional location reaction. After drying a permanent document is obtained which can be used for densitometric evaluation and which can be kept conveniently for future reference.

When current visualization methods are modified for the print technique, procedures in which the reagents are usually present in a single or in several sequentially applied solutions, it is essential to incorporate all reagents in sufficient amount into the paper in which the enzyme reaction proceeds. The concentration of substrates, coupling dyes, cofactors, etc., should be higher roughly by a factor of 5–10 in the solutions used for impregnating the paper than in those used for incubation of the compact gels for enzyme visualization. Sequential steps for enzyme visualization can be used but inclusion into the paper for printing of all reagents for staining should be tried whenever possible. When several steps are necessary for enzyme staining, the dipping technique (Smith, 1969) employing solutions of the reagent in highly volatile organic solvents is preferable to spraying the reagent, which latter procedure has frequently been found to result in diffuse zones.

2.   Advantages of the Print Technique

When used for enzyme detection, the print technique provides a very convenient and permanent document. The enzyme reaction is stopped on drying and for many enzyme visualization procedures little change is observed on storage of the dried print. Different detection methods can be applied in parallel on a single broad migration track. Enzyme visualization can thus be easily correlated with protein staining, as illustrated by Fig. 4 in which in the middle part the esterases of water-soluble barley proteins are shown

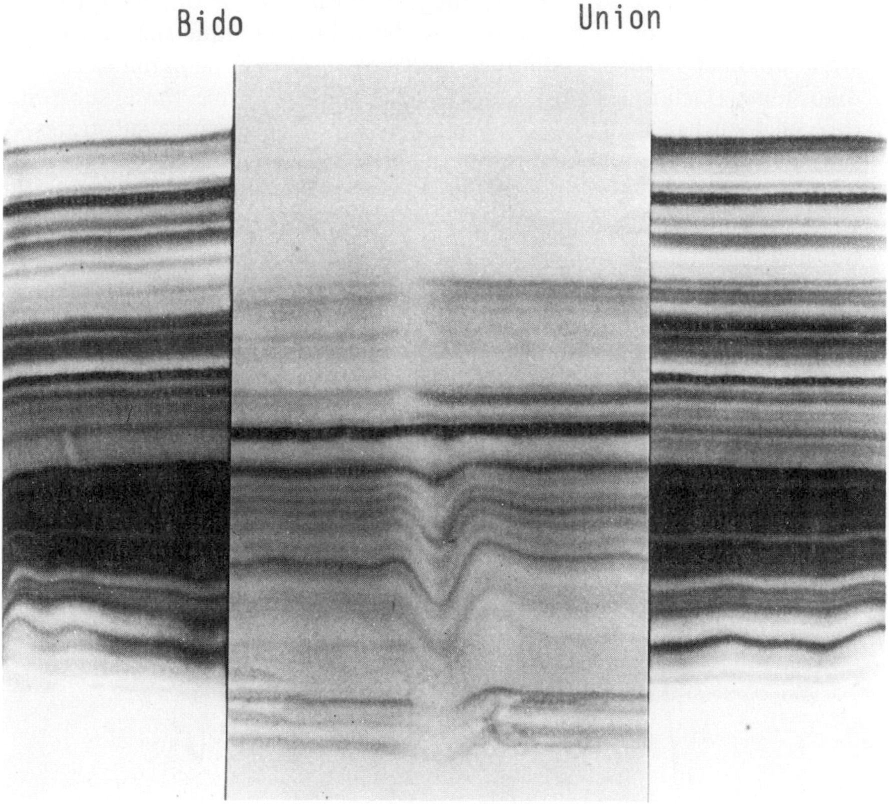

Fig. 4.   Thin-layer isoelectric focusing of water-soluble barley proteins in pH 3–10 carrier ampholytes. Proteins of two varieties (Bido and Union). Detection of esterases with β-naphthyl acetate and Fast Blue RR in the middle of the plate; proteins stained with Coomassie Brilliant Blue G-250 at both edges. [From Radola (1973a), used with permission of the New York Academy of Sciences.]

together with the protein staining at both edges. The sample was applied as a broad band and detection of enzyme activity and protein staining were thus possible on a single migration track, providing an excellent basis for correlating activity with protein distribution. On a sufficiently broad migration track glycoprotein or lipoprotein could be stained in addition to enzyme visualization and protein staining. Substrate specificity of an enzyme or group of isoenzymes can be studied by taking a print from a single migration track with papers impregnated with different substrates. In a similar way the pH optimum and the effect of inhibitors or activators, on multiple forms of an enzyme can be determined by printing with papers containing buffers of different pH values or the substance of interest.

High sensitivity is achieved by the print technique in enzyme visualization with some substrates. In the screening of 20 different secondary substrates for the visualization of peroxidase by the print technique the best results were obtained with o-toluidine, guaiacol, o-phenylenediamine, and o-dianisidine (Delincée and Radola, 1972). Sensitivity for these substrates differed by a factor of 10–100. With the two most sensitive substrates, o-phenylenediamine and o-dianisidine, as little as 0.001–0.01 μg of peroxidase could be detected. Versatility combined with high sensitivity is afforded by some fluorogenic substrates. To illustrate this visualization method, a pattern obtained after thin-layer isoelectric focusing of a commercial preparation of acidic phosphatases from potatoes is shown after activity detection with 4-methylumbelliferyl phosphate (Fig. 5). When increasing amounts of the enzyme are applied, enzyme zones can be detected in addition to the major components which in the densitometric tracing are visible as peaks

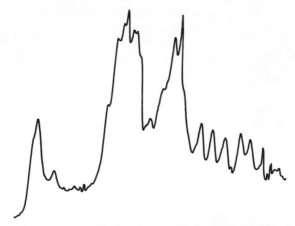

Fig. 5.   Detection of potato acid phosphatase with 4-methylumbelliferyl phosphate after thin-layer isoelectric focusing in pH 3–10 carrier ampholytes. Densitometry in reflectance at $\lambda_{ex} = 365$ nm and $\lambda_{em} = 650$ nm. Cathode at the right. [From Radola (1973a), used with permission of the New York Academy of Sciences.]

with only poor differentiation as a result of overloading. A great variety of fluorogenic substrates are already available and many new substrates can be expected to be introduced in the near future as a result of the rapid development of fluorometric essays for enzymes (Guilbaut, 1970).

An added advantage of the print technique is that any enzyme visualization reaction developed for thin-layer isoelectric focusing can be transferred with only minor modifications to detect enzyme activity following separation by other thin-layer methods employing granulated gels, such as thin-layer gel chromatography or thin-layer electrophoresis. Application of these methods, which are based on different separation principles, can be anticipated to further characterize physicochemically the analyzed enzyme.

3. Visualization of Enzymes Acting on Macromolecular Substrates

The print technique appears particularly useful for the visualization of enzymes acting on macromolecular substrates. With the techniques available for the visualization of these enzymes following isoelectric focusing in compact gels resolution is partially lost during incubation in the substrate solution (Vesterberg, 1973) or on diffusion of the enzyme into either a preformed gel layer (Vesterberg and Eriksson, 1972) or a gel overlayer (Wadström and Smyth, 1975). When using the print technique macromolecular substrates can be incorporated into the paper for the print as easily as low-molecular-weight substances. Since the substrate paper is in contact with the gel for only a few minutes and the reaction is stopped on drying the print, no resolution is lost on enzyme visualization. For a number of enzymes acting on macromolecular substrates negative staining techniques, with which absence of staining or a decreased intensity is indicative of enzyme activity, can be applied. Proteases represent a particularly important group of enzymes of this category (Radola, 1973a, 1975b). For detection of protease activity the paper for the print is impregnated with a buffered solution of, e.g., casein or hemoglobin or any other desired protein substrate. The paper is dried at an elevated temperature to denature the protein and to get better fixation of the protein substrate to the paper. After focusing a print is taken from the gel layer and proteolytic degradation is allowed to proceed in the paper at any desired temperature. The degradation products are washed out from the paper print with trichloroacetic acid and the undegraded protein in the print is subsequently stained, e.g., with Coomassie Violet R-150 or Light Green SF. White sharply focused zones of proteolytic activity appear on a uniformly stained background (Fig. 6). Application of a series of dilutions of the protease provides a rough estimate of the concentration of the individual proteolytically active components. A similar approach can be used for the detection of amylases, pectin depolymerase, pectin methylesterase, ribonuclease, etc. (Table III).

10% Pronase E   pH marker
40 µl   20 µl   10 µl   5 µl   proteins

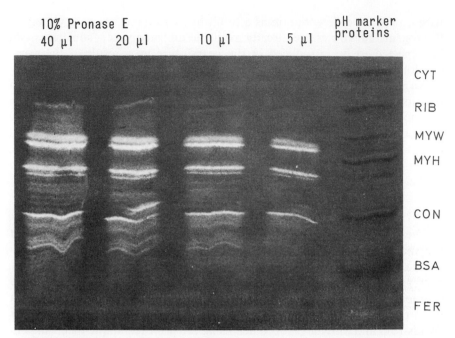

CYT

RIB

MYW

MYH

CON

BSA

FER

Fig. 6.   Detection of proteolytic activity in decreasing volumes of a 10% (w/v) solution of pronase E after thin-layer isoelectric focusing in pH 3–10 carrier ampholytes. An enzyme print was made with buffered paper (pH 7.2) impregnated with 0.3% hemoglobin. The background was stained with Coomassie Violet R-150, which also stained the pH marker proteins. CYT, cytochrome c, RIB, ribonuclease; MYW, sperm whale myoglobin; MYH, horse myoglobin; CON, conalbumin; BSA, bovine serum albumin; FER, ferritin. [From Radola (1975b).]

An alternative procedure with some enzymes acting on macromolecular substrates is to locate the low-molecular-weight products liberated by the enzyme. This procedure gives excellent results for many enzymes, e.g., amylases, cellulases, and glucoamylase (Table III). Detection of cellulase activity represents a particularly simple application because in this case the paper itself is the substrate. After degradation of the substrate in the paper the liberated reducing sugars are located in an additional step by means of a dipping reaction. Cellulase activity has been detected after thin-layer isoelectric focusing in a compact polyacrylamide gel by an indirect print technique (Eriksson and Pettersson, 1973). After focusing a viscous solution of carboxymethyl cellulose was sprayed on the gel, on top of which a chromatographic paper was subsequently placed. After contact for a few minutes the paper print was removed, dried, and sprayed to locate the degradation products. It follows from this application that after thin-layer

TABLE III

Enzyme Visualization by the Print Technique After Thin-Layer Isoelectric Focusing

Enzyme	Source	Visualization	Ref.
Peroxidase	Horseradish	Urea peroxide: o-toluidine, guaiacol	Delincée and Radola (1970, 1971, 1974a–c) Delincée et al. (1971a, b)
		Urea peroxide: 20 different secondary substrates	Delincée and Radola (1972)
	Barley, malt	Urea peroxide: guaiacol, o-toluidine/mesidine	Radola and Drawert (1970)
	Tobacco tissue cultures	Urea peroxide: guaiacol, o-toluidine/mesidine	Rücker and Radola (1971)
	Green beans	Urea peroxide: o-phenylenediamine	Delincée et al. (1974, 1975a)
	Sinapis alba Cicer arietum	Urea peroxide: o-dianisidine	Hösel et al. (1975)
Acid phosphatase	Potato	4-Methylumbelliferyl phosphate	Radola (1973a)
Esterase	Barley	β-Naphthyl acetate: Fast Blue RR	Radola (1973a) Drawert et al. (1973)
	Green beans	β-Naphthyl acetate: Fast Blue RR	Delincée et al. (1975a)
	Grapes	β-Naphthyl acetate: Fast Blue RR	Drawert and Müller (1973)
Lipoxygenase	Oats, soybeans	Linoleic acid: starch/KJ	Heimann et al. (1973)
	Green beans	Linoleic acid: o-dianisidine	Delincée et al. (1975a)
Polyphenoloxidase	Grapes	Mixture of polyphenols	Drawert and Müller (1973)
	Sunflower	4-Methylcatechol	Judel (1975)
Protease	Streptomyces griseus	Hemoglobin/casein: Light Green SF	Radola (1973a, 1975b)
	Bacteria, fungi	or Coomassie Violet R-150	Delincée (1975)
Pectin methylesterase	Tomato	Pectin: hydroxylammonium Chloride and $FeCl_3$	Delincée (1976)
Ribonuclease	Bovine pancreas	RNA: pyromine G	Delincée and Radola (1975b)
β-Amylase	Malt	Starch: aniline phosphate	Radola (unpublished)
Cellulase	Aspergillus	Filter paper: aniline phosphate	Radola (unpublished)
Glucoamylase	Aspergillus	Starch: aniline phosphate	Radola (unpublished)
Pectin depolymerase	Aspergillus	Pectin: ruthenium Red	Delincée (unpublished)

isoelectric focusing in compact polyacrylamide gels the print technique with either paper or cellulose acetate could also prove useful for the visualization of enzymes acting on macromolecular substrates.

4. Comparison with Other Separation Methods

On isoelectric focusing a high heterogeneity, superior to that obtained with other separation methods, was observed for many enzyme systems in some of the first applications of the method irrespective of anticonvective stabilization (Leaback and Rutter, 1968; Hayes and Wellner, 1969; Susor *et al.*, 1969; Delincée and Radola, 1970a). The results of isoelectric focusing were compared with those obtained with other separation methods for some enzymes of plant origin. When studied by thin-layer gel chromatography a single zone is observed for the peroxidase from green beans (Delincée *et al.*, 1975a) and also for horseradish peroxidase (Delincée and Radola, 1975a). More than 20 multiple forms, all with peroxidase activity, are detected for both peroxidases on thin-layer isoelectric focusing. The esterases of green beans show two zones of activity on thin-layer gel chromatography, five on thin-layer electrophoresis, and some 20 multiple forms with pI values distributed over a broad pH range on thin-layer isoelectric focusing. For many enzymes isoelectric focusing has disclosed an unprecedented heterogeneity. Despite the doubts expressed by some skeptics evidence is accumulating that the multiple forms found on isoelectric focusing represent physicochemical species and not methodological artifacts. By preparative isoelectric focusing some highly heterogeneous enzymes have been separated into individual components which were homogeneous on refocusing and could be analyzed by other physicochemical methods (Delincée and Radola, 1975a).

C. Two-Dimensional Separations

A combination of thin-layer isoelectric focusing with thin-layer gel chromatography has been described by Drawert and Müller (1971). A mixture of defined proteins has been separated by gel chromatography first according to size differences and in a second step according to charge differences. Thin-layer isoelectric focusing followed by polyacrylamide gel gradient electrophoresis has been applied to a study of serum proteins (Felgenhauer and Pak, 1975). Focusing in a layer of Sephadex G-75 was found useful because retardation of the high-molecular-weight proteins (e.g., α-macroglobulins, immunoglobulin M, and β-lipoprotein) could be avoided. Both two-dimensional methods provide "maps" which are more selective for protein characterization because the involved techniques depend on different principles of separation.

III. PREPARATIVE ISOELECTRIC FOCUSING

The development of efficient analytical separation methods is often followed by attempts to adapt the methods for preparative work. The many efforts aimed at scaling-up of polyacrylamide gel electrophoresis, one of the most successful analytical methods, reflect the difficulties in passing from analytical conditions to preparative scale work (Chrambach and Rodbard, 1971; Gordon, 1975). Although useful preparative separations can be achieved, the proliferation of devices indicates that preparative polyacrylamide electrophoresis is still in a stage of development. Preparative electrophoresis in blocks and columns (Bloemendal, 1963), employing a variety of different supports, continues to be used occasionally. While some of the techniques of preparative electrophoresis have a considerable loading capacity, their resolving power generally does not exceed that obtainable by conventional electrophoretic methods on nonsieving supports, and is rather poor as judged by application of high-resolution analytical methods such as polyacrylamide gel electrophoresis or isoelectric focusing.

Despite its great potential for the isolation and purification of amphoteric substances, which was recognized early (Vesterberg, 1968; Rilbe, 1970), preparative isoelectric focusing was not given sufficient consideration in the past. Different approaches have been suggested for preparative isoelectric focusing, including vertical columns with density gradient stabilization (Vesterberg, 1968, 1971b; Paléus et al., 1971), compact polyacrylamide gels in the form of either flat beds (Leaback and Rutter, 1968; Graesslin et al., 1972; Graesslin and Weise, 1974) or cylindrical gels (Finlayson and Chrambach, 1971; Righetti and Drysdale, 1973; Suzuki et al., 1973), zone convection focusing (Valmet, 1969), a polyethylene tubing (Macko and Stegemann, 1970), multi-membrane devices (Rilbe, 1970), and continuous-flow systems with gel or density gradient stabilization (Fawcett, 1973).

The two main methods of anticonvective stabilization used at present are density gradients and gels. Use of a density gradient for preparative separations in vertical columns has several limitations: loss of resolution due to remixing of proteins on elution from the column (cf. Sections I.A and II.A.1), low load capacity (Vesterberg, 1971b), sensitivity to local temperature fluctuations (Haglund, 1971), difficulties in ensuring efficient cooling which imposes finite limits on the column design (Rilbe, 1970), and isoelectric precipitation (Rilbe, 1970). Some of these drawbacks can be overcome by short vertical columns with either radial or axial cooling (Rilbe, 1973; Rilbe and Pettersson, 1975). The latter system particularly is capable of scale-up, and offers high load capacity as well as high total loads, but the resolution achieved in situ can be expected to be lost to an undefined degree because of diffusion after switching off the current and on elution of the column content.

For preparative isoelectric focusing in compact polyacrylamide gels a high load capacity, in terms of the amount of protein per square centimeter of a single zone, has been demonstrated for the cylindrical gels (Finlayson and Chrambach, 1971). It was claimed that up to 200 mg of protein can be processed without loss of resolution by increasing the gel volume (Righetti and Drysdale, 1973). The unfavorable geometry of the cylindrical gels puts a limit to a further increase of total load. Flat bed polyacrylamide gels were also used for preparative separations but little information is available on the load capacity of these systems. A common drawback of preparative iso-isoelectric focusing in compact polyacrylamide gels is the inconvenient elution of the separated proteins from the gel and their low recovery (Righetti and Drysdale, 1974). Quantitative elution by electrophoresis has been described for hemoglobins following preparative isoelectric focusing in cylindrical gels (Suzuki *et al.*, 1973).

1. Systems for Preparative Isoelectric Focusing in
   Horizontal Layers

When used for anticonvective stabilization of the pH gradient in preparative isoelectric focusing, horizontal layers of granulated gels offer a number of advantages (Radola, 1969, 1971, 1973a, 1975a,b). The geometry of a gel layer is favorable to heat dissipation, thus permitting the use of high field strengths, a factor important for resolution. The dimension of the gel layer can be varied with respect to length, width, and thickness, thus affording high versatility because both the separation distance and volume can be optimally adapted to the fractionation task.

Preparative isoelectric focusing can be performed in 0.06–0.1 cm layers, usually employed for analytical experiments by applying the sample as a broad band. Preparative separations are preferably carried out in gel layers of greater thickness. Figure 7 schematically represents the two available systems for

(a) small-scale separations on glass plates with gel layers up to 2 mm thick, and
(b) large-scale separations in troughs with gel layers up to 10 mm thick.

The gel volume in both systems in one of the commercially available apparatus with a 40×20 cm cooling area* ranges from 10 ml for a $20 \times 5 \times 0.1$ cm layer to 800 ml for the $40 \times 20 \times 1$ cm trough. Single proteins and protein mixtures at total protein loads between 0.025 and 10 g can be processed with excellent resolution in these systems. It seems feasible to scale-up the method further for even higher quantities of protein by increasing the

---

* Double Chamber, Desaga, Heidelberg, West Germany.

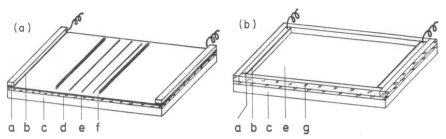

Fig. 7.   Preparative isoelectric focusing: (a) small-scale separations, (b) large-scale separations. a, electrode; b, filter paper pad soaked with electrode solution; c, cooling block; d, glass plate; e, gel layer; f, focused proteins; g, trough.   [From Radola (1975a).]

dimension of the gel layer. This could be achieved most simply by using broader layers at a constant separation distance and thickness of the layer. Longer separation distances could result in prohibitively long focusing times, whereas for layers thicker than 1 cm heat dissipation could prove to be the limiting factor.

2.   Resolution

Preparative isoelectric focusing on a quartz plate was performed with a mixture of proteins consisting of ovalbumin, β-lactoglobulin, myoglobin, ribonuclease, and cytochrome c (Radola, 1975b). Most of these proteins were previously separated by thin-layer isoelectric focusing (Radola, 1973b), thus facilitating comparison of preparative separations with analytical runs (cf. Figs. 1 and 3). Use of a quartz plate permitted direct densitometry of the focused proteins in the gel layer at 280 nm. Increasing amounts corresponding to 25, 50, 100, and 200 mg protein were applied on plates with 1- and 2-mm layers of Sephadex G-200. The densitograms convincingly demonstrate that the excellent resolution obtainable in analytical separations can also be achieved in preparative focusing, even at the highest protein load in these experiments (Fig. 8). The pattern for a number of the pH marker proteins used in thin-layer isoelectric focusing and the mixture in preparative experiments are very similar. For some of the proteins (e.g., ovalbumin, β-lactoglobulin, and horse myoglobin) the patterns of the preparative runs closely resemble those obtained by in situ scanning in analytical focusing systems with different anticonvective stabilization (Fredriksson, 1972; Catsimpoolas, 1973). These experiments demonstrate that it is possible to proceed from analytical to preparative separations without loss of resolution.

At the high protein load in these experiments and in preparative separations of many other proteins, sharply focused transparent zones could be seen directly in the gel layer when viewed against a dark background. These zones also prove that an excellent resolution is obtained in preparative iso-electric focusing. When preparative runs are evaluated by the print technique some of the peaks may appear insufficiently separated as a result of heavy background staining, which at the high protein loads cannot be completely destained, even if protein stains of low sensitivity (e.g., Light Green SF) are used.

The β-lactoglobulin was included to study the effect of isoelectric precipitation in preparative separations. Heavy bands of precipitated β-lactoglobulin

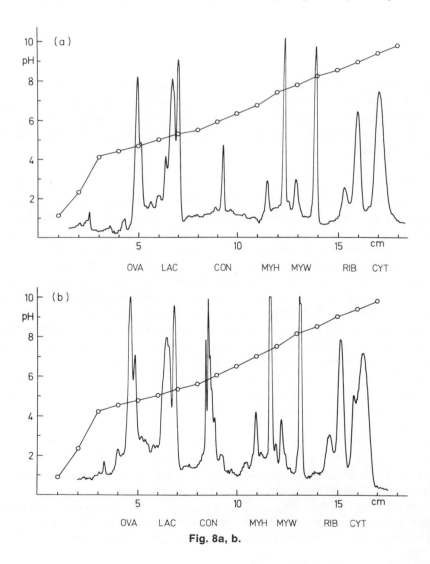

Fig. 8a, b.

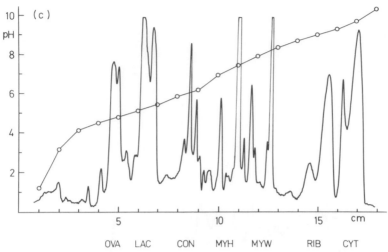

Fig. 8.   Focusing of a protein mixture on a 20 × 10 cm quartz plate. Sephadex G-200, pH 3–10 carrier ampholytes. Thickness of the gel layer and amount of total protein applied: (a) 1 mm, 25 mg; (b) 2 mm, 50 mg; and (c) 2 mm, 200 mg. Sample application in the middle of the plate. Focusing: 200 V for 16 hr followed by 600–800 V for 2–4 hr. Densitometry in transmission at 280 nm in the gel layer. ○, pH gradient. Proteins milligrams of each per milliliter: OVA ovalbumin (50 mg), crystallized once containing as impurity conalbumin (CON); LAC, β-lactoglobulin (50 mg); MYH, horse myoglobin (20 mg); MYW, sperm whale myoglobin (20 mg); RIB, ribonuclease (40 mg), crystallized five times: CYT, cytochrome c (20 mg). Major components of the proteins indicated in the patterns.   [From Radola (1975a).]

were observed in the gel layer during and after focusing but no adverse influence of the precipitated proteins on the resolution of the remaining proteins has been noticed. When working with granulated gels isoelectric precipitation is not a drawback, it could even facilitate the isolation of some proteins which after precipitation can easily be localized in the gel.

Great differences in the pI values for the proteins to be separated will facilitate preparative isoelectric focusing. From the experiments with direct densitometry in situ it can be inferred that in a pH 3–10 gradient with a 20-cm separation distance more than five components have been resolved per 1 pH unit, corresponding to a difference of less than 0.2 pH of the separated components. Resolution can be improved by working on a longer separation distance and by chosing narrow pH range carrier ampholytes. In preparative isoelectric focusing of sperm whale myoglobin on a 40-cm plate in pH 7–9 carrier ampholytes resolution approaches 0.05 pH (Fig. 9). The resolving power of isoelectric focusing has been claimed to be 0.01–0.02 pH (Vesterberg and Svensson, 1966) and a resolution close to this value was experimentally verified (Flatmark and Vesterberg, 1966; Vesterberg, 1967). It seems feasible to achieve this high resolution in preparative experiments also with the 40-cm

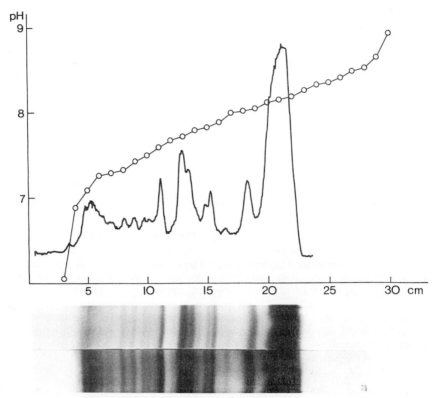

Fig. 9. Preparative isoelectric focusing of 0.5 g sperm whale myoglobin on a 40-cm plate. Sephadex G-200, pH 7–9 carrier ampholytes. Load capacity 2 mg/ml gel suspension. Focusing: 600 V for 16 hr followed by 800–1000 V for 8 hr. Bottom: densitogram of the unstained print; print stained with Light Green SF. ○, pH gradient. [From Radola (1975a).]

separation distance in still narrower pH carrier ampholytes than those commercially available. Narrow cuts of carrier ampholytes covering only 0.3–0.5 pH ranges can be easily obtained by preparative fractionation of 10–20% commercial carrier ampholytes in horizontal layers of granulated gels. A considerable improvement of resolution has been described for ovalbumin after fractionation of the commercial pH 4–6 carrier ampholytes (Radola, 1975a).

3. Load Capacity and Total Load

In order to compare isoelectric focusing in systems which employ a different geometry and/or other forms of anticonvective stabilization, load capacity is defined as the amount of protein milligrams per milliliter of gel suspension used for preparing the gel layer. The load capacity of a system is obtained by dividing the total protein load by the volume in which the pH gradient is established, which on focusing in layers of granulated gels corresponds roughly to the volume of gel suspension used for coating the plate. Depending on the experimental conditions of the preparative separations, the nominal pH gradient covered 70–90% of the total length of the gel layer. The appearance of a pattern of regular zones was used as the criterion in determining the highest permissible protein load capacity. Overloading resulted first in the formation of irregular zones, and a further increase in the amount of protein applied led to marked alterations in the pattern of zones and to the formation of droplets.

Single proteins as well as artificial and natural mixtures of proteins were employed to determine load capacity (Radola, 1975a,b). By thin-layer isoelectric focusing it was found that a limit of load capacity exists between 5 and 10 mg of total protein per milliliter of gel suspension. The validity of this finding was confirmed in small-scale preparative experiments on 20 × 10 cm plates coated with 50 ml of gel suspension. When load capacities for a protein mixture and single proteins are compared for the pH 3–10 and narrow pH range carrier ampholytes (Table IV) it becomes obvious that in addition to better resolution, the advantage of using narrow pH range carrier ampholytes is the decidedly higher load capacity. As a result of a nonuniformity of the pH gradient, which does not stabilize the focused proteins in all its parts equally well, some variation of the values for load capacity was observed for proteins with different isoelectric points.

Load capacity for a particular protein depends on the fractionation conditions (Table V). For total protein it is much higher when minor components are to be separated from an excess of major components. In experiments with partially purified egg white proteins, heavy overloading resulted for ovalbumin, the component present in excess, in distorted zones in the acidic range of the pH gradient. This overloading did not disturb focusing of conalbumin and lysozyme, the two minor components focused in the neutral and alkaline pH ranges.

TABLE IV

Load Capacity of Isoelectric Focusing in
Layers of Granulated Gels[a,b]

	pH 3–10	Narrow pH ranges
Protein mixture	5–10	5–10
Single proteins	0.25–1	2–4

[a] From Radola (1975a).
[b] Gels were Sephadex G-75 and G-200. Maximum load capacity: mg protein/ml gel suspension.

TABLE V

Different Fractionation Conditions in Preparative Isoelectric Focusing[a]

Egg white proteins,[b] partially purified	Maximum load capacity (mg protein per ml gel suspension used for preparing the layer)		
	Ovalbumin (77%)	Conalbumin (20%)	Lysozyme (3%)
	2–3	10	>10

[a] A: isolation of a major component (ovalbumin) in presence of minor components (conalbumin and lysozyme); or B: separation of minor components (conalbumin and lysozyme) from an excess of a major component (ovalbumin). From Radola (1975a).

[b] Percentage content based on electrophoresis.

[c] Focusing: pH 4–6 and pH 3–10 carrier ampholytes 2:1 v/v.

Based on these values for load capacity, gram amounts of pronase E, a highly heterogeneous proteolytic enzyme from *Streptomyces griseus*, were focused in the small- and large-scale systems described in Section III.1. In pH 3–10 carrier ampholytes 1.2 g pronase was focused in a 40 × 20 × 0.2 cm layer at a load capacity of 6 mg/ml. The pronase was resolved into more than 25 stainable protein zones with a resolution comparable to that of thin-layer isoelectric focusing (Radola, 1975a). In the 40 × 20 × 1 cm trough, 10 g of pronase was focused at a load capacity of 12 mg/ml (Fig. 10). Again an excellent resolution was observed in situ; many sharply focused zones could be seen directly in the gel layer. Also at this very high total protein load and load capacity the pattern closely resembled that of analytical separations.

The validity of the values determined for load capacity is further illustrated by Table VI in which data on preparative isoelectric focusing in horizontal layers of granulated gels are summarized for a number of single proteins and protein mixtures. The table reflects the versatility of the method with respect to the dimension of the gel layer and total load.

4. Recovery

Recovery in preparative isoelectric focusing will depend on a number of factors which are related either to the proper separation including elution from the gel or to additional steps necessary, for removal of the carrier ampholytes or concentration of the isolated fractions. Elution from granulated gels is simple, rapid, and quantitative. A loss of recovery at this step should be negligible in comparison with elution from compact polyacrylamide gels from which as a rule proteins cannot be eluted satisfactorily. The

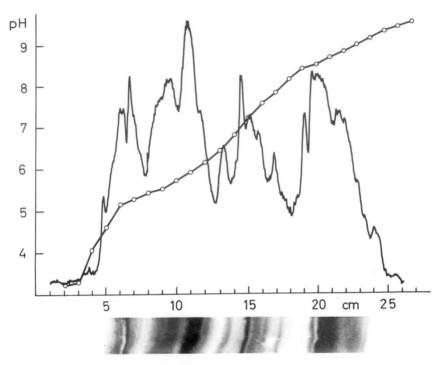

Fig. 10. Preparative isoelectric focusing of 10 g pronase E in a 40 × 20 cm trough. Thickness of the gel layer, 1 cm. Sephadex G-75, pH 3–10 carrier ampholytes. Load capacity, 12 mg protein/ml gel suspension. Focusing: 400 V for 16 hr followed by 800 V for 6 hr. Bottom: densitogram of the print stained with Light Green SF. ○, pH gradient. [From Radola (1975a).]

recovery of isoelectrically homogeneous proteins isolated by preparative focusing was studied in preparative refocusing experiments. Protein recoveries of 85–92% were found (Radola, 1973a). For crude protein mixtures recoveries of 80–90% were determined by eluting all proteins simultaneously from gel strips removed lengthways from the layer (Radola, 1975a). This approach gives a more reliable estimate of the recovery than the usual procedure in which protein recovery is calculated by summation of the protein content of individual isolated fractions. The latter method involves loss of the protein between the zones and the inevitable loss of some proteins resulting from manipulation of many small fractions. When this method was used to determine recovery of peroxidase after preparative focusing, about 75% of the enzyme activity could be recovered whereas the sum of activity of the individual zones gave only a recovery of 40–50% (Delincée and Radola, 1975a). The difference in enzyme recovery can be attributed to the loss of some minor enzyme zones and to zone cutting of the major zones.

TABLE VI   Preparative Isoelectric Focusing

Material fractionated	Gel	pH range	Dimensions of the gel layer (cm)	Total load (mg)	Load capacity (mg/ml)	Ref.
Horse myoglobin	Sephadex G-75	6–8	40 × 20 × 0.2	300	2	Radola (1969)
Sperm whale myoglobin	Sephadex G-75	7–9	40 × 20 × 1	1000	1.5	Radola (1971)
α_1 acute phase globulins	Sephadex G-75	5–7	34 × 16 × 0.4	20–30	0.08–0.12	Gordon and Dykes (1972)
Sperm whale myoglobin	Sephadex G-75	6–9	20 × 20 × 0.5	800	4	LaGow and Parkhurst (1972)
Nitrogenous substances from beer	Sephadex G-75	3–10	20 × 20 × 0.1	300	7.5	Wenn (1972a, b)
Human somatomedin	Sephadex G-75	3–10	46 × 8 × 0.4	100–200	1.3	Uthne (1973)
Human leucocytic pyrogens	Sephadex G-75	3–10	20 × 10 × 0.2	—	—	Dinarello et al. (1974)
Human chorionic gonadotropin	Sephadex G-75	3–10	40 × 20 × 0.75	—	—	Merz et al. (1974)
Horseradish peroxidase	Sephadex G-75	3–10, 3–5, 6–8, 7–9	40 × 20 × 0.2	300–400	2	Delincée and Radola (1975a)
Ribonuclease	Sephadex G-75	7–10	20 × 20 × 0.2	500	6.25	Delincée and Radola (1975b)
Chromatin nonhistone proteins	Sephadex G-50, 8 M urea	3.5–10	20 × 20 × 0.1	2.5	0.06	MacGillivray and Rickwood (1975)
Beef heart myoglobin	Sephadex G-75	6–8	20 × 3 × 0.4	100	4	Parkhurst and LaGow (1975)
Protein mixtures,	Sephadex G-75	3–10	20 × 10 × 0.1–0.2	25–10,000	1.25–12.5	Radola (1975a)
Sperm whale myoglobin,	Sephadex G-150	7–9	40 × 20 × 0.2			
Pronase E	Sephadex G-200 Bio-Gel P-200		40 × 20 × 1			
Normal adult human hemoglobin	Sephadex G-100	6–8	20 × 20 × 0.5			Steinmeier and Parkhurst (1975)
Canine secretory component	Sephadex G-75	5–7	20 × 10 × 0.06	100	6	Thompson et al. (1975)
Rabbit IgG immunoglobulin	Sephadex G-75	5–8	20 × 10 × 0.2	30	0.75	Urbain et al. (1975)

Determination of total activity in an aliquot portion of the gel layer is therefore a means to check inactivation inherent to the separation process.

Employing the same approach a recovery of 14–80% was determined for the proteolytic activity of pronase E after preparative isoelectric focusing. In this case enzyme recovery depended strongly on load capacity. Small-scale preparative experiments at load capacities ranging from 0.5 to 10 mg of pronase per milliliter of gel suspension have shown that the recovery of proteolytic activity increases with increasing load capacity from 14 to 80%. The highest recovery was repeatedly found at load capacities of 5–10 mg of pronase per milliliter of gel suspension. Since the caseinolytic activity of pronase is inhibited by EDTA, inactivation of pronase following isoelectric focusing could be due to the chelating properties of the carrier ampholytes (Davies, 1970). The dependence of enzyme recovery on load capacity and therefore also on the ratio of enzyme to the carrier ampholytes is compatible with this explanation. In contrast to observations describing reactivation of enzymes by adding metals necessary for activity after isoelectric focusing (Wadström and Smyth, 1975), addition of calcium or mixtures of different metal salts after focusing did not improve the recovery of pronase. The high load capacity of preparative isoelectric focusing is thus interesting not only with respect to economical considerations of this separation process. High ratios of proteins to carrier ampholytes improve the recovery of enzyme activity in those instances in which an inactivation is likely to occur because of the chelating activity of the carrier ampholytes.

Exposure to extreme pH values during focusing, particularly at low pH, is potentially detrimental to many proteins and could result either in artifactual patterns or low recoveries due to enzyme inactivation, or both. To avoid contact with extreme pH values the sample should be applied at a sufficient distance from the electrodes. It is well documented that during the initial stage of focusing, e.g., in pH 3–10 carrier ampholytes, extreme pH values are measured in the vicinity of the electrodes (Fig. 3a; Söderholm and Wadström, 1975). Samples applied too close to the electrodes would come in contact with the extreme pH values because the pH gradient is established at the electrode ends before the proteins can leave the application site. Components that are focused at extreme p*I* values should be protected from the detrimental effect of pH by efficient temperature control and short focusing time. The risk of denaturation can be at least partially avoided by establishing a pH gradient through prefocusing without applied sample. Contact of the sample could thus be substantially reduced, in some instances perhaps even at the expense of not reaching equilibrium conditions.

Dialysis, ultrafiltration, gel chromatography, ion exchange chromatography, precipitation with ammonium sulfate, and electrophoresis are some of the methods to remove the carrier ampholytes (Vesterberg, 1971b; Fawcett,

1975). The absence of sucrose or other solutes used in density gradients in the eluates from granulated gels is a definite advantage during later removal of the carrier ampholytes (Fawcett *et al.*, 1969). The efficiency of the methods for the removal of carrier ampholytes and their influence on protein recovery will strongly depend on both the properties of the material to be isolated and on the properties of the carrier ampholytes. Studies with ^{14}C-labeled Ampholine carrier ampholytes have shown that the highest apparent molecular weight is in the vicinity of 5000 daltons and that 0.75% has a molecular weight in the range of 1000–4000 daltons (Gasparić and Rosengren, 1975). In thin-layer gel chromatography of Ampholine carrier ampholytes two detection methods were applied: the caramelization reaction (Felgenhauer and Pak, 1973) and staining with Amido Black 10 B. Whereas the products giving the caramelization reaction were revealed after gel chromatography as a rather compact zone with low molecular weight, a diffuse but relatively intense staining with Amido Black 10 B was regularly observed in the extension of this compact zone into regions corresponding to components with molecular weights up to 5000–10,000 daltons. By contrast, no high-molecular-weight components could be detected by staining with Amido Black 10 B in the Servalyte carrier ampholytes (Radola, unpublished results). The presence of high-molecular-weight carrier ampholytes will cause difficulties on their removal by most of the methods indicated herein. Removal of these carrier ampholytes could prove particularly arduous in work with peptides and low-molecular-weight proteins.

5. Combination with Other Fractionation Methods

Although isoelectric focusing has excellent resolving power and isoelectrically homogeneous components can be isolated in a single step in high quantity, extensive purification of proteins usually will require the combined use of fractionation methods based on different properties of the separated molecules. The question then is raised at which step should preparative isoelectric focusing be used. The effect of different fractionation methods on the purity and homogeneity of horseradish peroxidase isoenzymes has been investigated (Delincée and Radola, 1975a). The horseradish peroxidase used as starting material, with an absorbance ratio $403/278 = 0.6$, consisted of more than 20 isoenzymes with pI values distributed over a wide range from pH 3.5 to 9.0 and numerous protein zones only partly paralleling the distribution of enzyme zones. Preparative isoelectric focusing of 100–400 mg of the crude peroxidase in pH 3–10 carrier ampholytes resulted in patterns essentially similar to those obtained in analytical focusing. Most of the isoenzymes present in sufficient quantity were isolated and proved on refocusing to be isoelectrically homogeneous, with only slight contamination with adjacent isoenzymes.

Despite the excellent separation of the individual isoenzymes their purity was not very high as judged by their absorbance ratio after preparative isoelectric focusing (Table VII). Preparative isoelectric focusing in narrow pH range carrier ampholytes slightly improved the purity of some of the major isoenzymes. When used as a single fractionation step gel chromatography brought about the same increase in absorbance ratio as preparative isoelectric focusing but no separation of individual isoenzymes. Cellulose ion exchange chromatography on microgranular CM or DEAE cellulose, used as a single fractionation step, was the only method to improve substantially the absorbance ratios for some of the fractions, which, however, were eluted as groups of isoenzymes with close isoelectric points. Isolation of individual

TABLE VII

Fractionation of Horseradish Peroxidase by Different Methods[a]

Fractionation method	Number of steps	Purity $(A_{403\ nm}/A_{275\ nm})$	Comments
Starting material	0	0.6	> 20 isoenzymes
Gel filtration	1	1.5	> 20 isoenzymes
Preparative isoelectric focusing, pH 3–10	1	1.7	Major isoenzymes isoelectrically homogeneous
Preparative isoelectric focusing, pH 6–8	1	2.3	Major isoenzymes isoelectrically homogeneous
CM cellulose	1	2.6	Mixture of isoenzymes
(1) CM cellulose (2) Preparative isoelectric focusing, pH 6–8	2	2.8	Major isoenzymes isoelectrically homogeneous
(1) Gel filtration (2) CM cellulose	2	3.0	Mixture of isoenzymes
(1) Gel filtration (2) CM cellulose (3) Preparative isoelectric focusing, pH 6–8	3	3.1–3.3	Isoenzymes isoelectrically homogeneous

[a] Based on Delincée and Radola (1975a).

isoenzymes and their extensive purification could thus not be achieved by any of the methods when used as a single fractionation step. Also, combinations of two of these methods were not effective in this respect.

In an attempt to keep the number of steps necessary to isolate individual isoenzymes with simultaneous high purification to a minimum, two fractionation sequences appear most attractive:

1.   column gel chromatography followed by chromatography on ion exchange cellulose and preparative isoelectric focusing, preferably in a narrow pH range carrier ampholytes, or

2.   chromatography on ion exchange cellulose followed by preparative isoelectric focusing and gel chromatography.

In both sequences the first fractionation step affords a high total load. With ion exchange chromatography the material is prefractionated with respect to the charge properties, thus providing the basis for use of narrow pH range carrier ampholytes in preparative isoelectric focusing which generally will enable more efficient separation than focusing in the broad pH range carrier ampholytes.

In experiments with horseradish peroxidase both fractionation sequences permitted the simultaneous isolation and purification of several isoenzymes. The yield and the absorbance ratios were higher in sequence 1 than in in sequence 2 which in the last step removed the carrier ampholytes also. An additional step, e.g. gel chromatography or ammonium sulfate precipitation followed by repeated washings, was necessary with sequence 2 for the removal of the carrier ampholytes. By both fractionation sequences individual isoenzymes with high absorbance ratios have been isolated, thus proving that methods based on different separation principles must be combined in order to obtain an efficient separation and purification. Of all the methods preparative isoelectric focusing was the only one by which individual isoenzymes could be isolated in a single step. Therefore, preparative isoelectric focusing as a single step could be the method of choice when the emphasis is on separation rather than purification. The purity of the isoelectrically isolated fractions will depend on the properties of the fractionated material and in particular on the distribution of the contaminating components in the isoelectric profile in the vicinity of the components of interest. In a recent study on the purification of a mammalian glucosaminidase similar conclusions for a generalized approach to protein isolation and purification were reached (Robinson and Leaback, 1974).

6.   Advantages

Preparative isoelectric focusing in horizontal layers of granulated gels is unique in that it evolves from analytical scale separations without changing the experimental setup. The scale of operation covers the range of comparatively low total loads for which a number of high-resolution electro-phoretic techniques are already available (Maurer, 1969; Chrambach and Rodbard, 1971; Gordon, 1975), and extends into a range of loads for which high-resolution techniques hitherto did not exist. Whereas for low protein loads other techniques of isoelectric focusing can be applied, the gram amounts of protein separated preparatively in layers of granulated gels re-present at the moment the upper limit of material fractionated by preparative isoelectric focusing. It is a feature of preparative isoelectric focusing in hori-zontal layers of granulated gels that the excellent resolution of analytical separations is attained in preparative applications under conditions of great experimental simplicity. The method has already been successfully used in a number of applications, as demonstrated in Table VI. Preparative isoelec-tric focusing in horizontal layers of granulated gels could prove highly efficient in many future applications.

Economical considerations are not a seriously limiting factor since, at the high load capacity which can and should be used, the cost of the carrier ampholytes will as a rule represent a minor fraction of the value of the usually already prefractionated material. However, application of the method could be further promoted if less expensive "industrial grade" carrier ampholytes would become available. Absence of the high-molecular-weight components in these carrier ampholytes could turnout to be more critical than some properties which are desirable in analytical work, e.g., when optical methods for evaluation are employed, but which are not further required in prepara-tive applications in which visualization is accomplished by means of protein or enzyme staining methods. While preparative isoelectric focusing has been little used so far, the development has now advanced to a stage at which the potential of the methods merits more attention as an efficient step to protein isolation and purification.

IV. APPLICATIONS

Thin-layer isoelectric focusing has been applied for the characterization of protein systems of plant and animal origin, and in a number of studies the results have been compared with those obtained by other separation methods. The general conclusion was reached in these studies that isoelectric focusing surpasses other techniques with respect to resolution. Some materials that previously resisted analysis by other methods, e.g., because of an excessive viscosity of the concentrated sample solution, could be fractionated by iso-electric focusing, which concentrates the individual components from dilute solutions into sharp zones. Water-soluble barley proteins were separated by thin-layer isoelectric focusing in layers of granulated gels and in density gradients (Radola and Drawert, 1970). This protein system was previously investigated by a variety of physicochemical methods, including such high-resolution techniques as polyacrylamide or starch gel electrophoresis and immunoelectrophoresis. The resolution obtained by thin-layer isoelectric focusing was better than that obtained by previous methods, more than 30 proteins being detected by staining. Barley and malt gave very similar patterns. Isoelectric focusing of beer proteins revealed the surprising fact that these proteins are physicochemically identical (with respect to their isoelectric points, zone banding, and relative distribution of the individual components) with the acidic proteins from barley and malt (Drawert et al., 1971, 1973). This finding together with previous results obtained with immunological methods (Grabar and Daussant, 1965; Nummi et al., 1969) strongly suggests that some of the barley proteins survive the technological sequence from barley to malt without change. Preparative isoelectric focusing in Sephadex G-75 layers was employed to fractionate the nitrogenous substances of beer and brewing material (Wenn, 1972a,b). The protein fraction stabilized the beer foam whereas other nitrogenous substances were ineffective in this respect.

Soluble proteins from grape juice of different varieties were analyzed by thin-layer isoelectric focusing and by different electrophoretic methods (Radola and Richter, 1972). A decidedly better resolution was achieved on thin-layer isoelectric focusing than on disk electrophoresis in polyacrylamide gels. Patterns very similar to those observed for the soluble proteins from juice were found for the proteins extracted from the acetone precipitated fraction of grape berries of different varieties (Drawert and Müller, 1973). Thin-layer isoelectric focusing was also used to study the soluble proteins from green beans (Delincée et al., 1975a), microorganisms (Drawert et al., 1973), and technical enzyme preparations following a radiation treatment to reduce microbial contamination (Delincée et al., 1975b). Sacroplasmic proteins from beef were resolved into 20–25 stainable protein zones (Radola, 1970, 1974). Irradiation of meat with ionizing radiation caused a decrease of the content of the most basic proteins, an effect that was more pronounced when the isolated sarcoplasmic proteins were irradiated in solution. Sarcoplasmic proteins isolated by different methods were compared for cod and trout (Bai and Radola, 1976). Snake venom of Crotalus atrox was studied by thin-layer isoelectric focusing (Radola, 1973b). The excellent resolution of the analytical experiments could also be achieved in preparative isoelectric focusing of the venom on a quartz plate (Radola, unpublished results). Ultraviolet densitometry of the proteins at 280 nm closely paralleled staining with Light Green SF. An electron-transferring flavoprotein and NADH dehydrogenase containing a novel orange flavin have been purified from the anaerobic bacterium Peptostreptococcus elsdenii. The isoelectric points of these proteins have been determined by thin-layer isoelectric focusing for the colored or fluorescent bands, the differences of the pI values being probably due to different proportions of their flavin chromophores (Whitfield and Mayhew, 1974a,b).

A number of applications refer to thin-layer isoelectric focusing in studies on enzymes. Commercial preparations of ribonuclease and desoxyribonuclease of different purity showed varying degrees of heterogeneity (Radola, 1973b). Protein staining revealed several components in ribonuclease preparations of the highest commercially available purity, and 8–15 components in desoxyribonuclease preparations. The method thus appears well suited to check the purity of commercial enzyme preparations. For some enzymes for which visualization reactions are not available the activity was determined in eluates from the gel, e.g., for tomato pectin methylesterase (Delincée and Radola, 1970b), for enzymes synthesizing UDP-apiose and UDP-D-xylose from UDP-D-glucuronic acid (Wellmann and Grisebach, 1971; Baron et al., 1972), an O-methyltransferase (Ebel et al., 1972), an enzyme catalyzing the transfer of D-glucose from UDP-D-glucose to flavonoids (Sutter et al., 1972), and enzymes involved in the biosynthesis of apiin (Ortmann et al., 1972). For cathepsin A from pig kidney, the activity was determined in eluates after thin-layer isoelectric focusing (Kawamura, 1974). Elution from gel segments after thin-layer isoelectric focusing was also used to detect and to determine the isoelectric point of a receptor protein from the epidermis cells of the pupae *Tenebrio molitor* (Schmialek et al., 1973), of penicillin-binding components in bacteria (Suginaka et al., 1972), and for an analysis of [^{3}H]oestradiol-17β binding proteins (Coffer and King, 1974).

Enzyme visualization has proved extremely useful in studies on the inactivation of horseradish peroxidase on heat treatment and irradiation. In view of the importance of the heat treatment for enzyme inactivation in food processing the effect of heat was studied with horseradish peroxidase as a model system (Delincée et al., 1971a) and for peroxidases of green beans (Delincée et al., 1974). Application of thin-layer isoelectric focusing in studies on the effect of ionizing radiation on horseradish peroxidase (Delincée et al., 1971b; Delincée and Radola, 1974a–c) and ribonuclease (Delincée and Radola, 1975b) provided much greater insight into the sequence of radiation-induced changes of the charge properties than previously available methods separating on the basis of charge differences. Further applications of the enzyme visualization technique are compiled in Table III.

Preparative isoelectric focusing was employed in a number of studies summarized in Table VI. Sperm whale myoglobin was separated into more than 15 components, and for the major components oxygen and CO ligand association kinetics were measured (LaGow and Parkhurst, 1972). The reaction of the isolated components with fluoride, azide, and cyanide were also studied (Parkhurst and LaGow, 1975). From fresh red cell hemolysate, five individual hemoglobin bands were isolated by preparative isoelectric focusing, column chromatography, or a combination of both methods (Steinmeier and Parkhurst, 1975). By preparative isoelectric focusing in Sephadex layers the individual hemoglobins were separated in the same order as found with analytical isoelectric focusing in cylindrical polyacrylamide gels (Drysdale *et al.*, 1971). Partial oxidation of the hemoglobin, described for polyacrylamide gels (Bunn and Drysdale, 1971), was not observed during focusing in the Sephadex layer. Excellent oxyhemoglobin samples were isolated which were spectroscopically indistinguishable from the unfractionated material. Components differing by as little as 0.11 pH unit were separated. The pI values measured in the gel layer agreed well with those obtained by other methods.

The α_1 acute phase globulins present in rat plasma were separated by preparative isoelectric focusing in Sephadex G-75 in the presence of 6 M urea (Gordon and Dykes, 1972). Single components were found on refocusing when individual zones isolated from the Sephadex layer were subjected to isoelectric focusing in flat polyacrylamide gels also in the presence of 6 M urea. Purified human chorionic gonadotropin was separated into six components by preparative isoelectric focusing in Sephadex layers (Brossmer *et al.*, 1971; Merz *et al.*, 1975a,b). The isolated fractions were homogeneous on refocusing in Sephadex but gave several bands when refocusing was performed in polyacrylamide gels in the absence or presence of 6 M urea. Individual bands of human liver α-L-fucosidase isolated by isoelectric focusing in a Sephadex layer focused as single components with identical pI values on refocusing in compact polyacrylamide gels (Thorpe and Robinson, 1975).

From circulating human white blood cells, pyrogens were isolated by preparative isoelectric focusing in Sephadex G-75 (Dinarello *et al.*, 1974). The isoelectric point of the neutrophil pyrogen was different from that obtained from monocytes. Both pyrogens differed also in molecular weight, precipitability with ethanol, and biological properties. Isoelectric focusing in 0.4-cm layers of Sephadex G-75 was employed as one of the purification steps in the large-scale production of somatomedin from human plasma (Uthne, 1973). A partially purified material gave only low recoveries when subjected to ion-exchange chromatography or ultrafiltration, but was recovered up to 70% after preparative isoelectric focusing. The factor responsible for the stimulation of DNA synthesis in human glia-like cells can be separated by isoelectric focusing from the factor stimulating DNA synthesis and sulfation in cartilage (Uthne, 1973).

Finally, an important application of analytical thin-layer isoelectric focusing is the detection of the patterns of carrier ampholytes by means of the caramelization reaction (Felgenhauer and Pak, 1973; Vinogradow *et al.*,

1973) or by staining with some protein dyes, e.g., Light Green SF (Radola, 1973b). In an extension of these studies an insight into the number and composition of different preparations of carrier ampholytes has been obtained by applying different detection methods in parallel (Radola, unpublished results). Preparative isoelectric focusing in open horizontal layers of granulated gels is well suited for the fractionation of carrier ampholytes at high concentrations of up to 10–20% (Radola, 1975a). The method can be applied equally well either for the purification of "homemade" carrier ampholytes or for the isolation of ampholytes of narrow pH ranges (covering as little as 0.2–0.3 pH unit) from commercial carrier ampholytes.

V. TECHNICAL APPENDIX

1.   Apparatus

The Double Chamber from Desaga (see footnote, p. 148) is best suited for isoelectric focusing in layers of granulated gels because it affords the highest flexibility. The apparatus is equipped with a 40 × 20 cm cooling block on which both analytical thin-layer plates as well as preparative plates and troughs with a length of up to 40 cm can be placed. The apparatus is equipped with adjustable platinum band electrodes which enable selection of any desired separation distance. Gas purging is possible, a feature important for alkaline range operation to exclude the adverse affect of carbon dioxide. Thin-layer isoelectric focusing can be performed also in the Multiphor LKB (Stockholm, Sweden), with only two fixed positions of the electrodes, and in the Electrophoresis System H-1000-D* with adjustable electrode distance. Adequate cooling, e.g., with a Lauda K2/RD refrigerated constant-temperature circulator, is recommended for focusing and pH measurements. Micro flat membrane electrodes with 1.5–2 mm diameter and a reference electrode are available.[†]

2.   Preparation of the Plate

Gel suspensions are prepared by suspending in a 1–1.5% (g/v) solution of the carrier ampholytes, e.g., 7 g of Sephadex G-75 or 4 g of Sephadex G-200, both "superfine." The required amount of dry gel for different products can be taken from Table I. The suitable gel consistency must be found by trial and error. The gel suspension is deaerated for a few minutes in a vacuum flask and can be used either immediately after preparation or after storage for periods up to several weeks at 4°C. For coating the plate an appropriate

---

[*] Serva, Heidelberg, West Germany.
[†] Desaga, Heidelberg or Ingold, Frankfurt, West Germany.

amount of gel suspension to give a final thickness of 0.06 cm is poured on a glass plate, e.g., 30 ml on a 20 × 20 cm plate. The gel suspension is spread on the plate by means of a glass rod. A perfectly even layer can be obtained by touching the plate from beneath with a vibrator or by repeatedly lifting one side of the plate several centimeters above the edge of the table and allowing it to fall. Gel suspensions of a correct consistency show a completely flat and shiny surface after this treatment. The gel plate is then dried in air until 1–3 mm fissures appear at its edges. This criterion, which is easily recognized after a little experience, corresponds to a water loss of 20–25%. Proper drying is essential for the quality of separations.

3. Sample Application

The sample can be applied directly on the surface of the gel layer as a spot with a micropipette, or as a band by placing the solution on the edge of a microscope cover slip (18 × 18 mm). Up to 50–75 μl can be applied per centimeter of the gel. A suitable applicator is available from Desaga. Broad starting zones are obtained by applying the samples with microscope slides or by means of rectangular glass pieces of suitable length. The samples should be applied at a certain distance from the electrodes, preferably not closer than 25% of the total separation distance, to avoid contact with extreme pH values at the beginning of focusing.

4. Focusing

Filter paper pads soaked with the electrode solutions, 0.5 M sulfuric acid at the anode and 2 M ethylenediamine at the cathode, are placed on both ends of the plate. Voltage is applied through platinum band electrodes put on the filter paper pads. A voltage gradient of ~10 V/cm is used until the current reaches approximately 20–25% of its initial value. The voltage is then raised to 20–40 V/cm. Focusing time depends on the length of the plate and on the pH of the carrier ampholytes. For a 20-cm separation distance 6–8 hr is adequate; 40-cm plates are run at the lower voltage overnight and then for several hours at the higher voltage.

5. Print Technique

The print is obtained by rolling onto the gel layer a dry chromatographic paper of suitable size and surface weight (~ 120 g/m^2). Care should be taken not to entrap air between the paper and the gel. After contact for 1–2 min the paper is removed from the gel and dried in a horizontal position in an oven at 110–115°C. Higher temperatures may cause difficulties in destaining the paper. The carrier ampholytes are washed out from the paper with 10% trichloroacetic acid. The proteins are preferably stained with Coomassie Brilliant Blue G-250 (Radola, 1973b).

For enzyme visualization the paper is impregnated with a nonvolatile buffer of suitable pH and buffering capacity (e.g., 0.3–0.5 M phosphate or Tris buffer). The dry, buffered paper is impregnated with the substrate dissolved in a volatile organic solvent, e.g., methanol or acetone. After short drying for solvent removal, the buffered, substrate-impregnated paper is rolled onto the gel layer. The enzyme reaction proceeds in the paper and enzyme bands are visible either immediately or after drying at any desired temperature.

6. pH Measurements

By means of the micro flat membrane electrodes the pH is measured directly in the gel layer, e.g., at 0.5–1 cm intervals. The pH should be measured along the migration track of the sample and not in an unloaded part of the gel. In work with colored proteins the pH is measured in the maximum of the protein concentration. Alternatively, the pH can be measured on 20 × 2 mm gel segments removed from the layer. A rough estimate of the pI values of an unknown sample can be obtained by running pH marker proteins on the same plate (Radola, 1973b).

7. Preparative Isoelectric Focusing

For preparative separations, plates with a thickness up to 2 mm can be obtained essentially as described for thin-layer focusing. Preparation of the layer is easier in troughs which are also commercially available. The gel in the plate or in the trough is dried until a layer is obtained which does not move when the plate is inclined at an angle of 45°. The sample is applied either directly on the surface of the gel layer or, with larger sample volumes, as a fairly liquid gel suspension into a slot. Voltage and time depend on the dimension of the gel layer. Charges up to 0.05–0.06 W/cm^2 are well tolerated by layers up to 1 cm thick. Proteins are located in the gel by a print with a heavy paper (250–500 g/m^2). Without printing, the proteins can be visualized as fluorescent bands by spraying the gel layer with a 0.003% solution of 8-anilino-1-napthalenesulfonic acid (Merz et al., 1975a). At high load capacities the proteins can be seen as transparent zones in the gel layer. On quartz plates detection at 280 nm is possible. After location in the layer, gel portions are removed and eluted by overlayering in a glass column, centrifugation, or pressing out the interstitial liquid with the aid of a syringe, followed by additional washing (Radola, 1975a).

REFERENCES

Arvidson, S., and Wadström, T. (1973). *Biochim. Biophys. Acta* **310**, 418–420.
Aspinall, G. O., and Cañas-Rodriguez, A. (1958). *J. Chem. Soc.* 4020–4027.
Awdeh, Z. L., Williamson, A. R., and Askonas, B. A. (1968). *Nature (London)* **219**, 66–67.

Bai, S. G., and Radola, B. J. (1976). *Chem. Technol. Mikrobiol. Lebensm.* (In press).

Baron, D., Wellmann, E., and Grisebach, H. (1972). *Biochim. Biophys. Acta* **258**, 310–318.

Bates, L. S., and Deyoe, C. W. (1972). *J. Chromatogr.* **73**, 296–297.

Bloemendal, M. (1963). "Zone Electrophoresis in Blocks and Columns." Elsevier, Amsterdam.

Bramhall, S., Noack, N., Wu, M., and Loewenberg, J. R. (1969). *Anal. Biochem.* **31**, 146–148.

Brewer, J. M. (1967). *Science* **156**, 356–357.

Brossmer, R., Merz, W. E., and Hilgenfeldt, U. (1971). *FEBS Lett.* **18**, 112–114.

Bunn, H. F., and Drysdale, J. W. (1971). *Biochim. Biophys. Acta* **229**, 51–57.

Carlström, A., and Vesterberg, O. (1967). *Acta Chem. Scand.* **21**, 271–278.

Catsimpoolas, N. (1968). *Anal. Biochem.* **26**, 480–482.

Catsimpoolas, N. (1969). *Sci. Tools* **16**, 1–5.

Catsimpoolas, N. (1971a). *Anal. Biochem.* **44**, 411–426.

Catsimpoolas, N. (1971b). *Separ. Sci.* **6**, 435–442.

Catsimpoolas, N. (1973). *Ann. N.Y. Acad. Sci.* **209**, 65–79.

Chrambach, A., and Rodbard, D. (1971). *Science* **172**, 440–451.

Chrambach, A., Reisfeld, R. A., Wyckoff, M., and Zaccari, J. (1967). *Anal. Biochem.* **20**, 150–154.

Clausen, J. (1969). *In* "Laboratory Techniques in Biochemistry and Molecular Biology" (T. S. Work and E. Work, eds.), Vol. 1, pp. 399–556. North-Holland Publ., Amsterdam

Coffer, A. I., and King, R. J. B. (1974). *Biochem. Soc. Trans.* **2**, 1269–1272.

Dale, G., and Latner, A. L. (1968). *Lancet* No. 7547, 848.

Davies, H. (1970). *In* "Protides of the Biological Fluids" (H. Peeters, ed.), Vol. 17, pp. 389–396. Pergamon, Oxford.

Delincée, H. (1976). *Phytochemistry* (In press).

Delincée, H. (Unpublished).

Delincée, H., and Radola, B. J. (1970a). *Biochim. Biophys. Acta* **200**, 404–407.

Delincée, H., and Radola, B. J. (1970b). *Biochim. Biophys. Acta* **214**, 178–189.

Delincée, H., and Radola, B. J. (1971). *In* "Protides of the Biological Fluids" (H. Peeters, ed.), Vol. 18, pp. 493–497. Pergamon, Oxford.

Delincée, H., and Radola, B. J. (1972). *Anal. Biochem.* **48**, 536–545.

Delincée, H., and Radola, B. J. (1974a). *Radiat. Environm. Biophys.* **11**, 213–218.

Delincée, H., and Radola, B. J. (1974b). *Radiat. Res.* **58**, 9–24.

Delincée, H., and Radola, B. J. (1974c). *Radiat. Res.* **59**, 572–584.

Delincée, H., and Radola, B. J. (1975a). *Eur. J. Biochem.* **52**, 321–330.

Delincée, H., and Radola, B. J. (1975b). *Int. J. Radiat. Biol.* **28**, 565–579.

Delincée, H., Radola, B. J., and Drawert, F. (1971a). *Experientia* **27**, 1265–1267.

Delincée, H., Radola, B. J. and Drawert, F. (1971b). *Int. J. Radiat. Biol.* **19**, 93–97.

Delincée, H., Becker, E., and Radola, B. J. (1974). *In Proc. Congr. Food Sci. Technol., 4th, Madrid* (In press).

Delincée, H., Becker, E., and Radola, B. J. (1975a). *Chem. Technol. Mikrobiol. Lebensm.* **4**, 120–128.

Delincée, H., Münzner, R., and Radola, B. J. (1975b). *Lebensm-Wiss. Technol.* **8**, 270–273.

Determann, H. (1969). "Gel Chromatography." Springer-Verlag, Berlin and New York.

Dinarello, C. A., Goldin, A. P., and Wolff, S. M. (1974). *J. Exptl. Med.* **139**, 1369–1381.

Drawert, F., and Müller, W. (1971). *Chromatographia* **4**, 23–26.

Drawert, F., and Müller, W. (1973) Lebensm. Unters. Forsch **153**, 204–212.

Drawert, F., Radola, B. J., Müller, W., and Görg, A. (1971). *Proc. Eur. Brew. Conv.* **13**, 479–488.

Drawert, F., Radola, B. J., Müller, W., Görg, A., and Bednař, J. (1973). *Proc. Eur. Brew. Conv.* **14**, 463–472.

Drysdale, J. W., Righetti, P. G., and Bunn, H. F. (1971). *Biochim. Biophys. Acta* **229**, 42–50.

Ebel, J., Hahlbrock, K., and Grisebach, H. (1972). *Biochim. Biophys. Acta* **269**, 313–326.

Eriksson, K. E., and Pettersson, B. (1973). *Anal. Biochem.* **56**, 618–620.

Fantes, K. H., and Furminger, I. G. S. (1967). *Nature (London)* **215**, 750–751.

Fawcett, J. S. (1968). *FEBS Lett.* **1**, 81–82.

Fawcett, J. S. (1970). *In* "Protides of the Biological Fluids" (H. Peeters, ed.), Vol. 17, pp. 409–412. Pergamon, Oxford.

Fawcett, J. S. (1973). *Ann. N.Y. Acad. Sci.* **209**, 112–125.

Fawcett, J. S. (1975). *In* "Isoelectric Focusing" (J. P. Arbuthnott and J. A. Beeley, eds.), pp. 23–43. Butterworths, London.

Fawcett, J. S., Dedman, M. L., and Morris, C.J.O.R. (1969). *FEBS Lett.* **3**, 250–252.

Fazekas de St. Groth, S., Webster, R. G., and Datyner, A. (1963). *Biochim. Biophys. Acta* **71**, 377–391.

Felgenhauer, K., and Pak, S. J. (1973). *Ann. N.Y. Acad. Sci.* **209**, 147–152.

Felgenhauer, K., and Pak, S. J. (1975). *In* "Progress in Isoelectric Focusing and Isotachophoresis" (P. G. Righetti, ed.), pp. 115–120. North-Holland Publ., Amsterdam.

Finlayson, G. R., and Chrambach, A. (1971). *Anal. Biochem.* **40**, 292–311.

Fischer, L. (1969). *In* "Laboratory Techniques in Biochemistry and Molecular Biology" (T. S. Work and E. Work, eds.), Vol. 1, pp. 151–396. North-Holland Publ., Amsterdam.

Flatmark, T., and Vesterberg, O. (1966). *Acta Chem. Scand.* **20**, 1497–1503.

Flodin, P. (1962). "Dextran Gels and Their Applications in Gel Filtration." Uppsala.

Fredriksson, S. (1972). *Anal. Biochem.* **50**, 575–585.

Gasparić, V., and Rosengren, Å. (1975). *In* "Isoelectric Focusing" (J. P. Arbuthnott and J. A. Beeley, eds.), pp. 178–181. Butterworths, London.

Godson, G. N. (1970). *Anal. Biochem.* **35**, 66–76.

Gordon, A. H. (1975). "Electrophoresis of Proteins in Polyacrylamide and Starch Gels." North-Holland Publ., Amsterdam.

Gordon, A. H., and Dykes, P. J. (1972). *Biochem. J.* **130**, 95–101.

Grabar, P., and Daussant, J. (1965). *Proc. Eur. Brew. Conv.* **10**, 147–155.

Graesslin, D., and Weise, H. C. (1974). *In* "Electrophoresis and Isoelectric Focusing in Polyacrylamide Gel" (R. C. Allen and H. R. Maurer, eds.), pp. 199–205. Gruyter, Berlin.

Graesslin, D., Trautwein, A., and Bettendorf, G. (1971). *J. Chromatogr.* **63**, 475–477.

Graesslin, D., Weise, H. C., and Czygan, P. J. (1972). *FEBS Lett.* **20**, 87–89.

Guilbaut, G. G. (1970). "Enzymatic Methods of Analysis." Pergamon, Oxford.

Haglund, H. (1971). *Methods Biochem. Anal.* **19**, 1–104.

Haglund, H. (1975). *In* "Isoelectric Focusing" (J. P. Arbuthnott and J. A. Beeley, eds.), pp. 3–22. Butterworths, London.

Haller, W. (1965). *Nature (London)* **206**, 693–697.

Harzer, K. (1970). *Z. Anal. Chem.* **252**, 170–174.

Hayes, M. B., and Wellner, D. (1969). *J. Biol. Chem.* **244**, 6636–6644.

Hedenskog, G. (1975). *J. Chromatogr.* **107**, 91–98.

Heimann, W., Dresen, P., and Schreier, P. (1973). *Z. Lebensm. Unters. Forsch.* **152**, 147–151.

Hinton, R. H., Burge, M. L. E., and Hartmann, G. C. (1969). *Anal. Biochem.* **29**, 248–256.

Hjertén, S. (1964). *Biochim. Biophys. Acta* **79**, 393–398.

Hösel, W., Frey, G., and Barz, W. (1975). *Phytochemistry* **14**, 417–422.

Johansson, B. G. (1971). *In* "New Techniques in Amino Acid, Peptide and Protein Analysis" (A. Niederwieser and A. Pataki, eds.), pp. 249–269. Ann Arbor Sci. Publ., Ann Arbor, Michigan.

Johansson, B. G., and Hjertén, S. (1974). *Anal. Biochem.* **59**, 200–213.

Jonsson, M., Pettersson, E., and Rilbe, H. (1969). *Acta Chem. Scand.* **23**, 1593–1599.

Jonsson, M., Pettersson, S., and Rilbe, H. (1973). *Anal. Biochem.* **51**, 557–576.

Judel, K. G. (1975). *Biochem. Physiol. Pflanz.* **167**, 233–241.

Kawamura, Y. (1974). *J. Biochem.* **76**, 915–924.

Koch, H. J. A., and Backx, J. (1969). *Sci. Tools* **16**, 44–47.

Korant, B. D., and Lonberg-Holm, K. (1974). *Anal. Biochem.* **59**, 75–82.

LaGow, J., and Parkhurst, J. J. (1972). *Biochemistry* **11**, 4520–4525.

Latner, A. L., and Skillen, A. W. (1968). "Isoenzymes in Biology and Medicine." Academic Press, New York.

Leaback, D. H., and Rutter, A. C. (1968). *Biochim. Biophys. Res. Commun.* **32**, 447–453.

Lewin, S. (1970). *Biochem. J.* **117**, 41 P.

MacGillivray, A. J., and Rickwood, D. (1975). *In* "Isoelectric Focusing" (J. P. Arbuthnott and J. A. Beeley, eds.), pp. 254–260. Butterworths, London.

Macko, V., and Stegemann, H. (1970). *Anal. Biochem.* **37**, 186–170.

Maurer, H. R. (1971). "Disc Electrophoresis and Related Techniques of Polyacrylamide Gel Electrophoresis." Gruyter, Berlin.

Merz, W. E., Hilgenfeldt, U., Dörner, M., and Brossmer, R. (1975a). *Hoppe-Seyler's Z. Physiol. Chem.* **355**, 1035–1045.

Merz, W. E., Hilgenfeldt, U., Brossmer, R., and Rehberger, G. (1975b). *Hoppe-Seyler's Z. Physiol. Chem.* **355**, 1046–1050.

Miles, L. E. M., Simmons, J. E., and Chrambach, A. (1972). *Anal. Biochem.* **49**, 109–117.

Nummi, M., Loisa, M., and Enari, T. M. (1969). *Proc. Eur. Brew. Conv.* **12**, 349–356.

Ortmann, R., Sutter, A., and Grisebach, H. (1972). *Biochim. Biophys. Acta* **289**, 293–302.

Peléus, S., Vesterberg, O., and Liljeqvist, G. (1971). *Comp. Biochem. Physiol.* **39B**, 551–557.

Parkhurst, L. J., and LaGow, J. (1975). *Biochemistry* **14**, 1200–1205.

Peterson, R. F. (1971). *J. Agr. Food Chem.* **19**, 585–599.

Quast, R. (1971). *J. Chromatogr.* **54**, 405–412.

Quast, R., and Vesterberg, O. (1968). *Acta Chem. Scand.* **22**, 1499–1508.

Radola, B. J. (1968). *J. Chromatogr.* **38**, 61–77.

Radola, B. J. (1969). *Biochim. Biophys. Acta* **194**, 335–338.

Radola, B. J. (1970). *In* "Colloquium on the Identification of Irradiated Foodstuffs," pp. 73–82. Euratom, Luxembourg.

Radola, B. J. (1971). *In* "Protides of the Biological Fluids" (H. Peeters, ed.), **18**, pp. 487–491. Pergamon, Oxford.

Radola, B. J. (1973a). *Ann. N.Y. Acad. Sci.* **209**, 127–143.

Radola, B. J. (1973b). *Biochim. Biophys. Acta* **295**, 412–428.

Radola, B. J. (1974). *In* "Identification of Irradiated Foodstuffs," pp. 27–43. Commission of the European Communities, EUR 5126, Luxembourg.

Radola, B. J. (1975a). *Biochim. Biophys. Acta* **386**, 181–195.

Radola, B. J. (1975b). *In* "Isoelectric Focusing" (J. P. Arbuthnott and J. A. Beeley, eds.), pp. 182–197. Butterworths, London.

Radola, B. J. (Unpublished).

Radola, B. J., and Delincée, H. (1971). *J. Chromatogr.* **61**, 365–369.

Radola, B. J., and Drawert, F. (1970). *Brauwissenschaft* **23**, 449–458.

Radola, B. J., and Richter, O. H. K. (1972). *Chem. Mikrobiol. Technol. Lebensm.* **1**, 41–50.

Righetti, P. G., and Drysdale, J. W. (1973). *Ann. N.Y. Acad. Sci.* **209**, 163–186.

Righetti, P. G., and Drysdale, J. W. (1974). *J. Chromatogr.* **88**, 271–321.

Rilbe, H. (1970). *In* "Protides of the Biological Fluids" (H. Peeters, ed.), Vol. 17, pp. 369–382. Pergamon, Oxford.

Rilbe, H. (1973). *Ann. N.Y. Acad. Sci.* **209**, 80–83.

Rilbe, H., and Pettersson, S. (1975). *In* "Isoelectric Focusing" (J. P. Arbuthnott and J. A. Beeley, eds.), pp. 44–57. Butterworths, London.

Riley, R. F., and Coleman, M. K. (1968). *J. Lab. Clin. Med.* **72**, 714–720.

Robinson, M. K., and Leaback, D. H. (1974). *Biochem. J.* **143**, 143–148.

Rodbard, D., Chrambach, A., and Weiss, G. H. (1974). *In* "Electrophoresis and Isoelectric Focusing in Polyacrylamide Gel" (R. C. Allen and H. R. Maurer, eds.), pp. 62–105. Gruyter, Berlin.

Rücker, W., and Radola, B. J. (1971). *Planta* **99**, 192–198.

Schmialek, P., Borowski, M., Geyer, A., Miosga, V., Mündel, M., Rosenberg, E., and Zapf, B. (1973). *Z. Naturforsch.* **28c**, 453–456.

Smith, I. (ed.) (1969). "Chromatographic and Electrophoretic Techniques," **2**, Heinemann, London.

Söderholm, J., and Wadström, T. (1975). *In* "Isoelectric Focusing" (J. P. Arbuthnott and J. A. Beeley, eds.), pp. 132–142. Butterworths, London.

Söderholm, J., Allestam, P., and Wadström, T. (1972). *FEBS Lett.* **24**, 89–92.

Steinmeier, R. C., and Parkhurst, L. J. (1975). *Biochemistry* **14**, 1564–1572.

Suginaka, H., Blumberg, P. M., and Strominger, J. L. (1972). *J. Biol. Chem.* **247**, 5279–5288.

Susor, W. A., Kochman, M., and Rutter, W. J. (1969). *Science* **165**, 1260–1262.

Sutter, A., Ortmann, R., and Grisebach, H. (1972). *Biochim. Biophys. Acta* **258**, 71–87.

Suzuki, T., Benesch, R. E., Yung, S., and Benesch, R. (1973). *Anal. Biochem.* **55**, 249–254.

Thompson, R. E., Reynolds, H. Y., and Waxdal, M. Z. (1975). *Biochem.* **14**, 2853–2859.

Thorpe, R., and Robinson, D. (1975). *FEBS Lett.* **54**, 89–92.

Urbain, J., Tasiaux, N., Leuwenkroon, R., Van Acker, A., and Mariame, B. (1975). *Eur. J. Biochem.* (In press).

Uthne, K. (1973). *Acta Endocrinol.* **73**, 1–35.

Valmet, E. (1969). *Sci. Tools* **16**, 8–13.

Vesterberg, O. (1967). *Acta Chem. Scand.* **21**, 206–216.

Vesterberg, O. (1968). Isoelectric Focusing of Proteins. Inaugural Dissertation, Karolinska Institutet Stockholm.

Vesterberg, O. (1969). *Acta Chem. Scand.* **23**, 2653–2666.

Vesterberg, O. (1971a). *Biochim. Biophys. Acta* **243**, 345–348.

Vesterberg, O. (1971b). *Methods Enzymol.* **22**, 389–412.

Vesterberg, O. (1973). *Acta Chem. Scand.* **27**, 2415–2420.

Vesterberg, O., and Eriksson, R. (1972). *Biochim. Biophys. Acta* **285**, 393–397.

Vesterberg, O., and Svensson, H. (1966). *Acta Chem. Scand.* **20**, 820–834.

Vinogradov, S. N., Lowenkron, S., Andonian, M. R., Bagshaw, J., Felgenhauer, K., and Pak, S. J. (1973). *Biochem. Biophys. Res. Commun.* **54**, 501–506.

Wadström, T., and Smyth, C. J. (1975). *In* "Isoelectric Focusing" (J. P. Arbuthnott and J. A. Beeley, eds.), pp. 152–177. Butterworths, London.

Weller, D. L., Meaney, A., and Sjörgren, R. E. (1968). *Biochim. Biophys. Acta* **168**, 576–579.

Wellmann, E., and Grisebach, H. (1971). *Biochim. Biophys. Acta* **235**, 389–397.

Wenn, R. V. (1972a). *J. Inst. Brew.* **78**, 377–383.

Wenn, R. V. (1972b). *J. Inst. Brew.* **78**, 404–406.

Whitfield, C. D., and Mayhew, S. G. (1974a). *J. Biol. Chem.* **249**, 2801–2810.

Whitfield, C. D., and Mayhew, S. G. (1974b). *J. Biol. Chem.* **249**, 2811–2815.

Wrigley, C. W. (1968). *J. Chromatogr.* **36**, 362–365.

Wrigley, C. W. (1971). *Methods Enzymol.* **22**, 559–564.

Zitko, V., and Bishop, C. T. (1966). *Can. J. Chem.* **44**, 1275–1282.

7 CONTINUOUS-FLOW ISOELECTRIC FOCUSING

John S. Fawcett

Department of Experimental Biochemistry
London Hospital Medical College
Queen Mary College, London, U.K.

I. INTRODUCTION

Isoelectric focusing is now well established as a high-resolution method for the analytical and preparative separation of proteins. Pioneered by Svensson for use with density gradient columns, the technique was soon adapted for use with polyacrylamide and Sephadex gels. It is now of interest to investigate whether isoelectric focusing is applicable to the fractionation of proteins on a large scale and, in particular, to establish whether the high degree of resolution associated with this technique can be maintained when scaling-up is attempted.

Of the preparative methods available, those using continuous-flow techniques appear the most attractive. They are capable of processing large quantities of material and, in addition, offer a number of advantages over other procedures. For example, continuous-flow methods are particularly suitable when toxic materials are involved, when sterile conditions are required, and when used in conjunction with other on-line separation processes.

In this chapter a number of different continuous-flow isoelectric focusing techniques are described and the relative merits and limitations of each method are discussed. Included in the final section are possible future developments for this fractionation procedure which could overcome some of the present limitations.

II. SCALING-UP OF ELECTROPHORETIC SEPARATIONS

For separation methods other than electrophoretic ones, a larger loading capacity can usually be obtained by increasing the size of the apparatus, although the gain in capacity is often at the expense of the resolution. For example, the scaling-up of column chromatographic separations can be achieved by a direct increase in column dimensions, but the degree of resolution can be maintained only if the uniformity of column packing, the efficiency of sample introduction, and the efficiency of the flow and removal of effluent all remain unaltered. With most electrophoretic methods, however, as a direct result of the electric current, the formation of heat within the system usually sets an upper limit to any increase in size. It becomes progressively more difficult to dissipate the heat as the size of the apparatus is increased.

This aspect is especially true for the density gradient column method as originally introduced for preparative isoelectric focusing. Any localized heating will cause convective mixing within the density gradient and destroy the stabilization. Since the LKB density gradient column* is cooled from both the inner and outer surfaces, the heat dissipation from it is very efficient. Nevertheless, when the voltage is applied, a small horizontal temperature gradient exists across the annular space with the warmest region approximately midway between the two cooled surfaces. Any extensive scaling-up by increasing the thickness of this annular space would ultimately give an excessive temperature rise resulting in an unacceptable level of convective mixing in this region.

Increased column capacity could be obtained by increasing the diameter of the annular space without altering the distance between the cooling surfaces. These wide annular columns would retain their cooling efficiency but the collection of focused zones at the base of the column would present major

---

* Produced by LKB, Bromma, Sweden.

difficulties. Even with density gradient columns in current use, narrowly separated zones are partially remixed as the gradient is funneled from the annular space into the narrow exit tube. This effect becomes progressively worse as the diameter of the column is increased.

These difficulties do not arise in the very short density gradient columns recently designed by Rilbe (1973) where the cooling is from the upper and lower surfaces. In these columns the efficiency of cooling is independent of the column width, and columns of very large capacity can be envisaged. The ingenious method of rotating the column through 90° from the vertical to the horizontal position redistributes the density gradient and gives a geometry which allows efficient collection of the separated zones.

The use of Sephadex gel as the stabilizing medium for preparative iso-electric focusing has been investigated in detail by Radola (Chapter 6). The capillary structure of the Sephadex bed is very effective in stabilizing focused zones against the effect of gravity. Convective mixing caused by localized heating or temperature gradients within the gel bed is almost totally eliminated. Sephadex-stabilized systems are therefore capable of high loading capacity and allow the use of relatively large bed volumes. Radola has fractionated 10 g of protein mixture by isoelectric focusing in a 1-cm-thick bed of Sephadex (bed volume 800 ml). This represents a major achievement in scaling-up and it is suggested (Radola, 1973a) that further increases in capacity are feasible with the possibility of fractionating 100 g of protein mixture.

Another method of preparative isoelectric focusing is that based on zone convection (Valmet, 1969) where the separation takes place in a horizontal corrugated channel. One of the virtues of this method is that proteins that precipitate at or near their isoelectric point are localized at the position of precipitation and do not spread and contaminate neighboring protein zones. Since this technique involves cooling from the upper and lower surfaces it too should be capable of scaling-up by increasing the width of the apparatus. The method is applicable to high loading capacity since regions of high density are confined within the trough and do not mix with adjacent compartments.

III. CONTINUOUS-FLOW ELECTROPHORESIS

In spite of the increased capacity envisaged for the various preparative methods listed in the preceding section it is likely that the processing of large quantities of protein mixture by these methods would require a number of consecutive batch experiments. It is in these circumstances that it would be advantageous to use continuous-flow methods especially if the apparatus could be operated automatically.

Over the past 25 years continuous-flow electrophoresis (also known as continuous-deflection electrophoresis) has attracted considerable attention as a means of scaling-up electrophoretic separations. First suggested by Philpot (1940), the technique consists of applying an electric field perpendicular to the direction of flow of a continuously moving stream of electrolyte (Fig. 1a). A solution of the sample is introduced continuously, at a fixed point, into the liquid stream, and any uncharged components of the sample move with the liquid flow. Charged components migrate in the electric field at rates determined by their electrophoretic mobilities and are deflected from the direction of liquid flow. Separated components can thus be collected continuously at the exit positions.

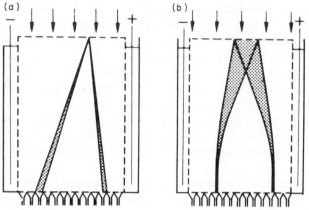

Fig. 1. Diagram illustrating the principle of (a) continuous-flow electrophoresis and (b) continuous-flow isoelectric focusing. Direction of liquid flow is indicated by arrows.

The first successful apparatus for continuous-flow electrophoresis were described by Svensson and Brattsten (1949) and by Grassmann and Hannig (1950a). In these designs a horizontal electric field was applied across the electrolyte which flowed continuously down a narrow rectangular trough packed with anticonvectant powder (glass beads, powdered glass, quartz sand, etc.). The main problem with this apparatus was the maintenance of uniform and parallel liquid flow through the packed bed for the duration of the experiment. In the original apparatus this was attempted by applying a constant hydrostatic pressure with the aid of Mariotte flasks but later designs (e.g., Brattsten, 1955) incorporated peristaltic pumps to control both inflow and outflow rates.

An alternative design was developed in which the migration occurs in a rectangular sheet of filter paper (Grassman and Hannig, 1950b; Brattsten and Nilsson, 1951; Durrum, 1951; Strain and Sullivan, 1951). With this apparatus the rate of flow of electrolyte is determined by the grade of filter paper selected; reasonably parallel flow is obtained with comparatively simple apparatus. However, unfavorable properties such as adsorptive and electroosmotic effects of the anticonvective material have severely limited the application of these techniques.

Undoubtedly the most successful design of continuous-flow electrophoresis equipment is based on the technique introduced by Barrollier *et al.* (1958) and developed extensively by the elegant work of Hannig (1961, 1969). Known as free-flow electrophoresis this technique consists of two parallel glass plates 50 × 50 cm spaced only 0.3–0.5 mm apart to form a very narrow separation chamber. Electrolyte flows through this chamber and electrophoresis is conducted in free solution. No anticonvectional packing is used to maintain stabilization. By positioning the plates almost horizontally thermal convection and gravitational mixing are virtually eliminated. High potential gradients may be used with this apparatus since very efficient cooling is obtained by blowing cold air onto the outer surfaces of the glass plates.

This apparatus has a good capacity (e.g., 40 ml serum per day) and has been widely used for the separation of proteins and nucleic acids. A large number of applications have been reported (Hannig, 1969).

The absence of a stabilizing medium in free-flow electrophoresis has allowed cells and cellular particles to be separated and has opened up a new area for electrophoretic purification. For these purposes it is necessary to use an apparatus with a vertical separation chamber, otherwise the particles sediment onto the horizontal surface of the lower cooling plate. However, in the vertical configuration, zones of increased density are now subjected to gravitational instability and, in addition, any unevenness in the cooling results in convectional mixing. It is therefore essential for the cooling to be rigorously controlled. In some apparatus this has been achieved by the use of Peltier cooling batteries, and commercial apparatus using this technique are currently available. The whole field of free-flow electrophoresis has been discussed in some excellent reviews by Hannig (1969, 1972).

A continuously flowing density gradient is another method for large-scale preparative electrophoresis. The technique was first suggested in the late 1930s by Philpot (1940) but at that time a number of practical difficulties prevented him from perfecting his apparatus. Almost 15 years later density gradient stabilization for electrophoresis in columns was introduced by Kolin (1954) and Svensson and Valmet (1955). Later, Mel (1960, 1964) adapted the column density gradient into a continuous-flow electrophoretic method.

Mel's apparatus consisted of a horizontal channel 30 cm long, 3 cm high, and 0.7 cm wide with membranes (dialysis tubing) separating the top and bottom of the channel from electrode compartments. Electrolyte solution in a range of sucrose concentrations is fed continuously into one end of the channel via 12 entrance tubes. It flows horizontally through the apparatus at right angles to the applied electric field and out through the exit tubes.

The technique has been used for the fractionation of human serum lipoproteins by Tippetts (1965) and in the purification of vaccinia virus by Robinson (1973) who was able to process the viral suspension at the rate of 500 ml/hr in an apparatus 60 cm long.

Mel reported that it was possible to establish reproducible and uniform pH gradients by using different buffers; Robinson has also used shallow pH gradients (pH 6.0–8.0) and streams of different buffer concentrations. However, the separations obtained by both groups were due to differences in electrophoretic mobilities and were not based on the principles of isoelectric focusing.

The concept of continuous-flow isoelectric focusing was introduced by Durrum and Pickels (1958). In a patent describing equipment for continuous-flow electrophoresis they suggested means of introducing appropriate buffers to provide a pH gradient for the separation and concentration of components on the basis of their isoelectric points.

The use of carrier ampholytes to produce a more stable pH gradient in continuous-flow isoelectric focusing was discussed in a lecture given by Rilbe in 1969 at a colloquium on the protides of the biological fluids at Bruges (Rilbe, 1970), but Seiler et al. (1970a), using their own version of the free-flow apparatus, were the first to demonstrate the application of this technique to the fractionation of proteins. More recently this laboratory (Just et al., 1973, 1975a,b) has extended this method to include the fractionation of cells and cell particles. Other reports have appeared describing the separation of proteins by continuous-flow isoelectric focusing, using the free-flow apparatus (Prusik, 1974), vertical troughs packed with Sephadex (Fawcett, 1973; Hedenskog, 1975), and with a continuously flowing density gradient (Fawcett, 1973).

These different approaches to continuous-flow isoelectric focusing: free-flow through a narrow trough, flow through a Sephadex packed bed, and a continuously flowing density gradient, are discussed in detail in Section V.

IV. ADVANTAGES OF CONTINUOUS-FLOW ISOELECTRIC FOCUSING

The principle of continuous-flow electrophoresis and continuous-flow isoelectric focusing is illustrated in Fig. 1. With continuous-flow electrophoresis a zone will continue to emerge at the same exit position provided the ratio of electrolyte flow to electrophoretic migration remains constant throughout the run. For a constant current it is essential, therefore, that the liquid flow remain uniform and at a constant rate to maintain a steady-state condition during the experiment. Successful applications of this technique are governed by the extent to which these conditions can be maintained.

With continuous-flow isoelectric focusing the situation is entirely different. Once the amphoteric components have migrated to their isoelectric points they have zero electrophoretic mobility and will then move only in the direction of the liquid flow (Fig. 1b). Thus, small changes in the rate of liquid flow have no effect on the exit positions of focused zones. In addition, the focusing mechanism will compensate for small irregularities in the pattern of liquid flow, since the position of the focused zone is determined by the pH gradient established between the electrode solutions. Furthermore, the pH gradients obtained from carrier ampholytes are not changed appreciably by small variations in field strength or temperature. Thus the method has built-in stability and the position where focused components emerge from the exit ports will remain constant for the duration of an experiment. Elaborate apparatus designs or special precautions to prevent drifting of the focused zones are unnecessary. It is only when gross changes occur in flow rate or applied potential that a slight shift of the focused positions is observed. This is caused by a slow drift of the pH gradient and is discussed in Section VI.C.

As with all other isoelectric focusing methods, the width of the sample zone in continuous-flow isoelectric focusing is unimportant. The sample may be applied as a very broad zone or, if necessary, can be spread over the whole width of the separation cell as is the custom with the column density gradient apparatus. Thus dilute solutions of protein mixtures may be fractionated, often eliminating the necessity of a concentration procedure.

In the electric field the amphoteric components will migrate to their isoelectric point where they will concentrate to narrow bands and, as they flow through the apparatus, the width of the bands will be maintained by the focusing mechanism. By contrast, when continuous-flow electrophoresis is used, the sample must be applied as a very narrow zone in the electrolyte stream and the width of the separated zones will spread by diffusion as the zones flow through the apparatus.

One very important feature of the continuous-flow isoelectric focusing technique is that the sharpness of the focused zones can be maintained by the applied potential right up to the time at which the zones enter the exit ports. This ensures that the high degree of resolution obtained within the separation cell is maintained in the isolated fractions. This procedure is not possible with column density gradient methods. When emptying the column, the separated zones begin to spread by diffusion as soon as the current is switched off, and further disturbance of the zone occurs as they pass down the column and into the fraction collector. The effect is seen when colored proteins are focused in the density gradient column, and is especially pronounced with zones focused to very narrow bands by application of high electric potentials.

After isoelectric focusing in Sephadex, using Radola's procedure, zone spreading before the fractions are isolated is of less importance. Diffusion is reduced by the Sephadex medium, but, if the position of the focused zone is detected by the paper print method, a delay of up to 30 min often occurs before the bed is cut into segments; during this period separation of closely focused zones deteriorates.

In the Valmet zone convection apparatus, zone spreading during the isolation of fractions can be completely eliminated. When the lid is removed immediately after the current is switched off, the liquid level is lowered so that fractions are isolated in each of the compartments in the trough.

A further advantage of the continuous-flow technique is that the sample can be introduced into any region of a preformed pH gradient. This is achieved simply by positioning the sample feed downstream at a point where the pH gradient is already established. A sample introduced at a pH region near its isoelectric point will reach the focused position more rapidly, thus allowing a greater through-put and reducing inactivation of labile material.

V. EXPERIMENTAL METHODS

A. Free-Flow Isoelectric Focusing

1. Vertical Type

The apparatus used by Seiler *et al.* (1970b) has a nonplanar separation cell and is operated in the vertical position. The separation cell illustrated in Fig. 2 is a U-shaped channel formed by fixing Lucite blocks (A and B) around three sides of a central Teflon-coated metal cooling plate (C). An electrode block (E) containing electrode chambers (F) is positioned at the back of the cooling plate. The complete assembly is attached together with tension bolts (G) and forms a compact unit. In the front block (A) and the electrode block (E) the bolt holes are enlarged, and with the aid of different spacer plates (M) it is possible to vary the thickness of the separating channel. The cell is 36 cm long and has an effective width of 11 cm; the thickness of the channel is normally 0.05 cm.

This novel design affords easy assembly of the electrode compartments; the semipermeable membrane (L) is firmly held in position by the silicone

Fig. 2. Vertical free-flow apparatus of Seiler *et al.* (1970b) used for continuous-flow iso-electric focusing. (a) Cross section through the separation cell. I (to left of broken line), section at height of tension bolts; II section at height of ampholyte supply tubes. A, separation cell front block; B, separation cell side block; C, cooling block; D, cooling liquid channel; E, electrode block; F, electrode channel with electrodes; G, tension bolt; H, cooling block locating screw; K, silicone gasket; L, electrode membrane; M, spacer; N, electrode solution exit tube; O, ampholyte supply tubes. (b) Photograph of complete unit. Front view of separation cell, carrier ampholyte reservoir at top, sample injection pump at left, and 48-fold pumping unit and fraction collector at base.

gasket (K), forming a leak-free seal. Electrode solution, stored in 2-liter reservoirs, is continuously pumped through the electrode cells during separation: 1% phosphoric acid or 5% acetic acid for the anode and 1.5% ethanolamine for the cathode solution.

Ampholyte solution, stored in a cooled reservoir at the top of the apparatus, is supplied to the separation cell at a constant hydrostatic head by a
number of entry tubes around the top of the front and side blocks. Additional
tubes located in these Lucite blocks allow the introduction of the sample
at various positions in the cell. Fractions are collected via 48 exit tubes
arranged around the three sides of the base.

Cooling liquid, maintained at a temperature of $1 \pm 0.05°C$, is circulated
through the cooling block, which is made of aluminum and insulated by a
thin Teflon film. This allows very efficient cooling of the liquid stream;
average field strengths of up to 125 V/cm* can be applied. Temperature
changes in the liquid stream are monitored by a thermistor located within
the separation cell.

Some of the results obtained from the isoelectric focusing of albumin
and gamma globulin in this apparatus are shown in Fig. 3 (Seiler *et al.*,
1970a). Separations of bovine gamma globulin in pH 3–10 Ampholine and
at an average field strength of 90 V/cm are shown for two different flow
rates. Resolution was improved by the longer residence time (100 min),
suggesting that the shorter period (50 min) was insufficient for optimal
focusing.

Seiler *et al.* (1970a) have also fractionated horse myoglobin and bovine
hemoglobin in a narrow range Ampholine (pH 6–8). At an average field
strength of 90 V/cm and residence time of 100 min the focused zones are
not as sharp as expected. Possibly, owing to slower electrophoretic mobilities
in the shallow pH gradient, these proteins have insufficient time to attain
optimal focusing.

Using the same apparatus, Just *et al.* (1973, 1975a) have demonstrated
the application of free-flow isoelectric focusing to the fractionation of mixed
red blood cells of different species. Figure 4a shows the separation obtained
with a mixture of human, rabbit, and mouse red blood cells in pH 3–10
Ampholine and at an average field strength of 110 V/cm. By injecting the
cells at a point halfway down the apparatus, at a position where the pH
gradient is already partly established, the residence time is reduced to a
minimum and morphological damage is prevented.

Although clearly fractionated in the separation chamber, the isoelectric
points of the different red blood cells are very close, causing a slight overlap
of the zones in the collecting tubes. When the same mixture was focused
in a narrow range Ampholine gradient (pH 5–7) a better separation was
obtained (Fig. 4b) and the different species were collected as homogeneous
fractions.

---

* Since the conductance varies along the pH gradient all field strengths quoted represent
average values.

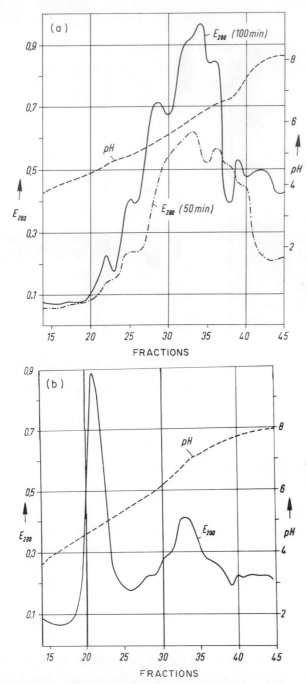

Fig. 3. Continuous-flow isoelectric focusing of proteins in the free-flow apparatus shown in Fig. 2 using Ampholine (pH 3–10) and a field strength of 90 V/cm. (a) Bovine gamma globulin: -·-, residence time 50 min ;——, residence time 100 min (protein concentration doubled). (b) Bovine albumin, residence time 100 min. [From Seiler *et al.* (1970a).]

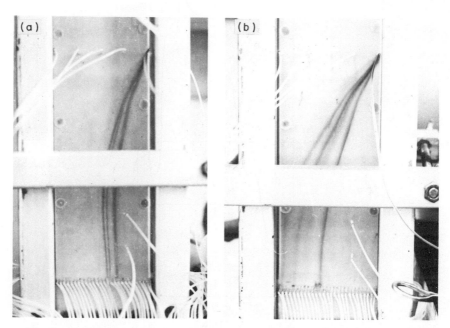

Fig. 4.   Separation of a mixture of red blood cells from human (left zone), mouse (middle
zone), and rabbit (right zone): (a) in 1% Ampholine pH 3–10; (b) in 1% Ampholine pH 5–7.
Residence time for cells 7 min. Field strength 110 V/cm. [From Just *et al.* (1975a).]

This group has also given a preliminary report (Just *et al.*, 1973, 1975b)
on the continuous-flow isoelectric focusing of the light mitochondrial
fraction obtained from rat liver. However, electron micrographs of the
fractions and distribution of a number of marker enzymes indicated that
the separation was incomplete.

2.   Horizontal Type

The continuous free-flow apparatus used by Prusík (1974) for isoelectric
focusing is a more conventional design. His instrument, similar to that
originally designed by Hannig (1961) but modified for isoelectric focusing,
is shown in diagramatic form in Fig. 5. Carrier ampholyte solution is
pumped through a horizontally oriented separation cell to give a liquid
film 0.05 cm thick, 50 cm wide, and with an effective length of 44 cm. The
apparatus is cooled by a current of cold air and is capable of dissipating
$0.36 \ W/cm^2$.

Prusík first fractionates his carrier ampholytes in a preliminary run
through the apparatus. The six input reservoirs are then filled with the
appropriate pooled fractions, one of which also contains the protein mixture.
By introducing the prefractionated ampholytes in this way the pH gradient

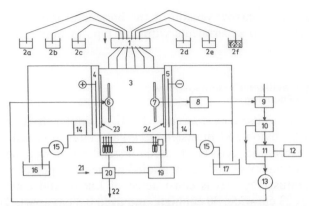

Fig. 5. Schematic diagram of continuous-flow isoelectric focusing with horizontal free-flow apparatus. 1, pump; 2, reservoirs of prefractionated carrier ampholyte (protein sample added to 2f); 3, separation chamber; 4, 5, electrode chambers; 6, air-cooling inlet; 7, air-cooling outlet; 8, temperature detector; 9, pneumatic amplifier; 10, proportional mixer; 11, heat exchanger; 12, cooling unit; 13, fan; 14, constant-level device; 15 centrifugal pump; 16, acid reservoir; 17, base reservoir; 18, fraction collector; 19, high-frequency-level detector; 20, air valve; 21, atmospheric air inlet; 22, vacuum; 23, 24, ion exchange membranes. [From Prusík (1974).]

is established much more rapidly, thus reducing the residence time. Furthermore, by this method the electrical conductance across the inflow end is lowered and less heat is produced in this section. Nevertheless, Prusík reports that the separation chamber must be located horizontally in order to limit thermal convections caused by uneven development of Joule heat (Section VI.B).

Prusík uses a wide separation cell (50 cm); thus, because of the increased migration distance, a longer time is required to establish the focused condition. In experiments with pig pancreatic amylase using Ampholine pH 5–8 (0.3%) the residence time was 150 min and the applied potential 60 V/cm.

Apparatus for free-flow electrophoresis is available commercially in a variety of different models: a wide version for use in a horizontal plane and both wide and narrow versions for vertical use. It would seem that when used for isoelectric focusing they require only minor alteration. Of course, it is necessary to have independent circulation for the acid and base electrode solutions, and with some units it may be necessary to provide a slower rate of pumping to the separation chamber in order to obtain optimal focusing. It would be an advantage to alter the electrodes so that a lower potential is applied across the inflow end to give a more even Joule heating (Section VI.B). The long migration distance of the wider models is a disadvantage when used for isoelectric focusing. For the fractionation of soluble components that do not precipitate during isoelectric focusing it would seem that a narrow version of the horizontal model would be the more suitable but no experiments with this type of apparatus have been reported.

B. Capillary-Stabilized Continuous-Flow Isoelectric Focusing

The equipment used by Fawcett (1973) is similar to that originally described by Svensson and Brattsten (1949) and later modified by Winsten *et al.* (1963), except that glass beads used for stabilization have been replaced by beads of Sephadex G-100 or by graded particles of polyacrylamide gel.

The apparatus (Fig. 6a) is constructed of Lucite and consists of two cooling plates 23 cm wide, 30 cm high, and fixed 0.3 cm apart. To facilitate good heat transfer the cooling surface is constructed from 2-mm-thick Lucite sheeting, strengthened by small spacer blocks arranged at 5-cm intervals. Although a thin glass cooling surface would be preferable, Lucite was chosen for ease of construction.

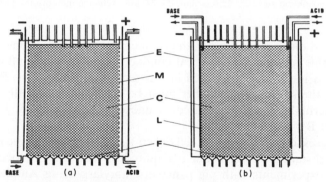

Fig. 6. Capillary-stabilized continuous-flow isoelectric focusing apparatus. (a) Vertical section of apparatus with semipermeable membrane (M) separating the cell (C) from the electrode (E). Hatched area is filled with Sephadex or polyacrylamide beads supported on filter membrane (F). (b) Apparatus without semipermeable membrane with electrode solution flowing down the end of the cell alongside a Lucite barrier (L) to form an electrical conducting layer with the electrodes. Only 12 of the 54 outflow tubes are illustrated. [From Fawcett (1973), used with permission of the New York Academy of Sciences.]

The electrode vessels are separated from the trough by semipermeable membranes (M), formed by polymerizing polyacrylamide gel within the pores of porous polyethylene sheeting. When dry these membranes can be glued into position, thereby simplifying apparatus construction. The dried material is regenerated by immersion in water for 24 hr to regain its original structure. Membranes of this type have the advantage of being tough and rigid, have negligible electroosmotic flow, and may be dried and regenerated by soaking many times without any detrimental effect on the membrane structure.

The separation cell (C) is packed with preswollen Sephadex beads which are supported on a filter (F) above the outflow tubes. Ampholyte is supplied to the cell via a constant head device or by a multichannel pump with an overflow arrangement. The flow through the cell is controlled by a variable-speed multichannel pump unit; with the aid of additional pumps, the anode and cathode compartments are slowly perfused with solutions of acid and base, respectively. The sample, dissolved in carrier ampholyte, is continuously injected just under the surface of the gel bed, often at a number of different positions.

When a voltage is applied to the apparatus there is some transfer of liquid through the membrane into the cathode compartment which interferes with compounds focusing at the basic end of the pH gradient. This has been prevented by the construction of a modified apparatus (Fig. 6b). In this unit the membranes have been omitted and replaced by a plastic barrier (L) across most of the section except for a region at the lower end where a nylon mesh filter (F), usually 2–5 cm in length, separates the cell from the electrode compartment. During focusing acid solution is continuously injected just under the top layer of the Sephadex bed adjacent to the plastic barrier by the anode. This solution will flow down the edge of the bed as a narrow zone, and, because of its high conductance, will form an electrical conducting layer with the electrode. Approximately two-thirds of the flow of the acid solution is continuously siphoned off at the top of the anode compartment, while the remainder of the acid solution will continue to flow down the bed and be drained by the end outflow tubes. In a similar manner base solution is continuously injected into the top of the Sephadex bed at a corresponding position at the cathode end.

In addition to eliminating membranes, a further advantage of this apparatus design is that it allows for a more uniform distribution of Joule heat. By suitable adjustment of the concentration of the acid and base solutions and by controlling the width of these conducting layers, a reduced electric potential is applied across the top of the separation cell. This procedure prevents excessive formation of heat at the top of the cell where the unfocused ampholytes have a high conductance (Section VI.B).

A photograph of the complete apparatus (Fig. 7) shows hemoglobin focused to its isoelectric point when halfway through the apparatus. With soluble proteins the unit maintains constant focusing conditions over long periods but substances that precipitate within the separation cell and clog the packing cause serious disturbances to liquid flow and to the separated zones.

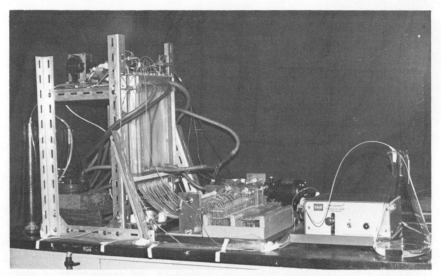

Fig. 7. Photograph of continuous-flow isoelectric focusing in Sephadex-stabilized medium. View of complete apparatus with pump units and collecting tubes. Hemoglobin introduced into two positions is focused when approximately halfway through the Sephadex bed. [From Fawcett (1973), used with permission of the New York Academy of Sciences.]

A smaller version of this apparatus has been used by Gianazza *et al.* (1975) for the fractionation of carrier ampholytes. Using 8% ampholyte solution and a flow rate of 20 ml/hr the equipment was operated continuously for 10 days. The pH gradient remained virtually constant during this period.

In a study of the maximum separation capacity of the Sephadex stabilized equipment, Hedenskog (1975) has constructed a larger apparatus with a glass separation cell for improved cooling efficiency. With a cell volume of 345 ml this apparatus has a capacity of 150 mg protein/hr when used with a potential of 50 V/cm.

The choice of media for these capillary-stabilized systems has not been fully investigated. In general the stabilizing properties would be expected to improve with a decrease in particle size and increase in porosity, but other factors, including resistance to liquid flow, adsorption and filtration of insoluble material, and electroosmotic flow, must all be taken into consideration.

Preliminary experiments using fine glass beads gave unsatisfactory results, presumably because of electroosmotic flow. No improvement was obtained from either extensive cleaning or by siliconization of the glass beads.

Various grades of Sephadex have been studied and it has been found that Sephadex G-75 and G-100 both give satisfactory results. In agreement with results reported by Radola (1973b) with Sephadex slab experiments, grades of Sephadex less porous than G-50 are unsatisfactory because of electro-osmosis possibly caused by acid groups attached to the dextran matrix. The less porous grades of Sephadex contain more dextran per unit volume and will thus have higher concentrations of these groups.

C. Continuous-Flow Density Gradient Isoelectric Focusing

1. Apparatus with Long Migration Path

Fawcett (1973) constructed a modified form of Mel's apparatus for continuous-flow isoelectric focusing. In its original form the unit (made of Lucite) consisted of two cooling plates assembled to form a separation cell 23 cm high, 35 cm long, and 0.3 cm wide (Fig. 8). Fifty-four tubes evenly spaced along one end distribute the density gradient; the horizontal flow is maintained by controlling both the inflow and outflow with a 108-channel peristaltic pump.

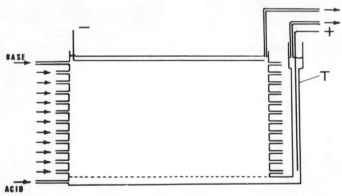

Fig. 8. Vertical section of continuous-flow density gradient apparatus. Fifty-four tubes (only 12 illustrated) introduce the sucrose gradient on the left and a corresponding number of tubes remove fractions at the opposite end. Dense electrode solution flows along the bottom section forming an electrical conducting layer with the electrode tube (T). [From Fawcett (1973), used with permission of the New York Academy of Sciences.]

The main divergence from Mel's design is the arrangement of the electrodes, since isoelectric focusing requires only small volumes of electrode solutions. The top membrane is discarded and the cathode electrode (stainless steel wire gauze) dips directly into the separation cell. The bottom electrode compartment consists of a channel that runs along the base of the apparatus and is connected to the external tube (T) containing the platinum anode. Dense sucrose solution containing orthophosphoric acid is pumped through the electrode channel and out through the top of the tube (T). By controlling the concentration of acid circulating through the electrode channel a reduced electric potential is applied across the entry end of the separation cell, giving a more uniform production of Joule heat (Section VI.B).

The presence of a membrane along the base of the separation cell is optional, but it does ensure better long-term stability of the pH gradient in the lower section of the cell.

Human hemoglobin, when injected at three different positions into this apparatus, migrated to a focused position when approximately halfway through the cell (Fig. 9). In this experiment 1% Ampholine (pH 3–10) was used and the residence time was 14 hr. The applied potential was 300 V across the entry end, increasing to 1200 V at the outflow end. A slight shift of the focused zone toward the top (cathode) electrode can be seen as it flows through the apparatus; this shift is caused by the slow drift of the pH gradient (Section VI.C).

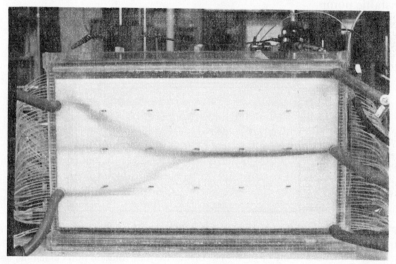

Fig. 9. Photograph of the isoelectric focusing of hemoglobin in a continuous-flow density gradient apparatus. Protein solution injected at three positions on the left migrates to the focused position when approximately halfway through the cell. Experimental conditions: 1% Ampholine (pH 3–10); applied potential, 300 (inflow end) to 1200 V (outflow end); residence time 14 hr. Cathode electrode along the top. [From Fawcett (1973), used with permission of the New York Academy of Sciences.]

At constant voltage and residence time the long-term reproducibility obtained with this equipment is excellent. If precipitation does not interfere with the fractionation, zones focus at the same position and are collected in the same fraction tube day after day. Some typical results of the fractionation of ox pituitary proteins are shown in Fig. 10.

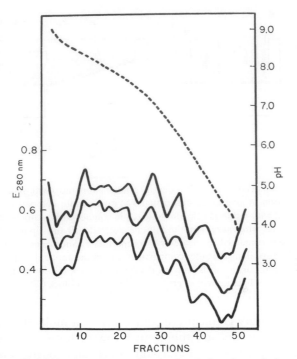

Fig. 10. Reproducibility of continuous-flow isoelectric focusing technique. Separation of bovine pituitary proteins in a continuous-flowing density gradient with a pH 3–10 gradient. Bottom curve: separation after 24 hr. Separations after 48 and 72 hr are displaced upward on the graph by 0.1 and 0.2 OD unit, respectively.

2.   Apparatus with Short Migration Path

Recently Fawcett (unpublished) has constructed a continuous-flow density gradient apparatus suitable for large-scale isoelectric focusing. It has a short migration path and is cooled from the bottom surface only—two of the suggestions originally proposed by Philpot (1940). This design allows minimum residence time, and the vertical temperature gradient, formed as a result of the bottom cooling, strengthens the density gradient stabilization.

The apparatus shown diagramatically in Fig. 11 consists of a narrow Lucite frame (A) 70 cm long divided into three compartments by the strips (B). The lower edge of (B) is tapered and is fixed 1.5 mm above the base of the apparatus which is a thin plastic film (C). Efficient cooling is obtained by ensuring good thermal contact between this plastic film base and a horizontal aluminum cooling block (D).

The two outer compartments contain the dense electrode solution and the center compartment contains the density gradient. Both the inflow and

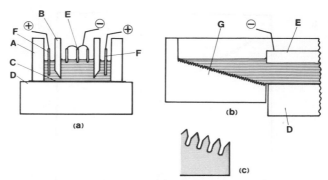

Fig. 11. Diagram of continuous-flow density gradient apparatus with short migration path. (a) Transverse section of long horizontal trough (A) divided into compartments by partition strips (B) and fitted with a thin plastic film base (C) in contact with a cooling block (D). Long strips of stainless steel wire gauze form the cathode electrode (E) and dip into the top layers of the density gradient contained in the central compartment. Short lengths of platinum wire gauze arranged in pairs along the outer compartments form the anode electrodes (F). (b) Longitudinal section of end portion with grooved distribution block (G) for entry and removal of the density gradient. (c) Enlargement of grooves in distribution block (G).

outflow of the density gradient are controlled by a multichannel peristaltic pump connected by capillary tubing to a row of holes arranged along an inclined plane at each end of the trough. Each tube locates with the lower section of one of the grooves in the distribution block (G) which allows the smooth entry and removal of liquid layers 0.25 mm thick.

Dense sucrose solution containing 1% phosphoric acid flows along the bottom of the apparatus to form an electrical conducting layer 1.5 mm thick. In the center compartment a steep sucrose density gradient containing ampholyte flows above this conducting layer to form a liquid column 1 cm high. On top of this an upper layer (2–3 mm) of cathode solution (1% TEMED or 1% ethanolamine) flows along the surface. The cathode electrode (E), consisting of vertical strips of stainless steel wire gauze, dips into the top layer. Three lengths of platinum wire gauze (10 cm long) positioned at equal intervals along the side of each outer compartment form the anode electrodes (F). Maximum potential is applied to the electrode pair adjacent to the outflow end, whereas the potential applied to the center pair is reduced to approximately half the voltage with a further reduction of voltage applied to the electrodes at the inflow end.

The delivery tubing between the pump unit and the trough passes through a cooling bath maintained at 1°C so that a precooled density gradient is introduced into the apparatus. This initial cooling together with the cooling obtained via the thin plastic base allows focusing with potentials of up to 80 V/cm and a residence time of only 8 min. However, with these conditions, the temperature of the upper layers of the density gradient rises to 26°C when using 1% Ampholine (pH 3–10). For separating heat-labile substances it is necessary to operate at lower potentials and consequently longer residence times. For example, when hemoglobins and pituitary proteins are focused in 1% Ampholine (pH 3–10) with an applied potential of 40 V/cm and residence time of 15 min, the temperature of the upper layers of the density gradient rises to only 16°C.

The unit is capable of fractionating 1–2 g of protein mixture per hour; this capacity could be increased by using a wider apparatus. Components of mixtures focus to very narrow zones and the resolution obtained from the separation of test mixtures is almost as good as that obtained by the long-migration-path apparatus.

D. Other Apparatus Possibly Suitable for Continuous-Flow Isoelectric Focusing

1. Multimembrane Compartment Cells

Much of the early pioneering work on isoelectric focusing used multi-compartment cells in which the cells were separated by membranes to prevent convectional mixing (Williams and Waterman, 1929). Considerable difficulties with this method were reported by Rilbe (1970) of which the main one was the electroosmotic flow caused by the membrane material. The low ionic strength at the focused state results in extensive flow which is intensified by any charged material precipitated on the membrane surface.

If the problems of electroosmosis can be overcome, this method can be adapted to a continuous-flow technique, and equipment with very large capacity can be envisaged. The compartments, which are normally fitted with cooling coils, could easily be increased in size without altering the migration distance. For continuous-flow application each compartment could be fitted with a "serpentine" partition, complete with inflow and outflow facilities, so that a continuous stream of solution could slowly circulate past each membrane. Apparatus of this type would be useful for preliminary separation, but any precipitation at the isoelectric point would obviously impose serious limitations.

2. Valmet Type Zone Convection Apparatus

A very wide version of this zone convection apparatus (Valmet 1969), but with a short electrophoretic migration distance, could be adapted for continuous-flow operation, with liquid flowing along the grooves of the apparatus and at right angles to the applied potential. This would represent a version of the horizontal free-flow apparatus but modified by having ridged cooling plates. If positioned to give a slight downward inclination of the liquid flow, it could operate with a high loading capacity and be suitable for preliminary separation. It could not be expected to be capable of high resolution.

3.   Philpot's Continuous-Flow Rotating Separator

An ingenious apparatus designed by Philpot (1965, 1973) for large-scale continuous electrophoretic separation should be applicable to isoelectric focusing. It is claimed to have a very high capacity capable of fractionating 1–1.5 liters of sample solution per hour.

The space for electrophoretic separation is a narrow (5 mm) annulus between a stationary central cylindrical electrode and a rotating outer cylindrical electrode assembly, both equipped with electrodes and semipermeable membranes. A laminar flow of solution is pumped continuously through the annular space in an upward direction and stabilization is maintained by a gradient of angular velocity resulting from the rotation of the outer electrode. The applied potential across the narrow annulus separates the components into concentric zones. At the top of the apparatus the flow is deflected through 90° and the separated zones are collected continuously via 30 outlets.

In a preliminary account (Thomson *et al.*, 1973) the application of this technique to the electrophoretic fractionation of a variety of biological fluids has been given, but as yet no application to isoelectric focusing has been reported. If the degree of stabilization obtained by this method is adequate, this equipment with its short migration path and large through-put could prove extremely valuable for very large-scale isoelectric focusing. Construction of the apparatus is difficult, requiring precision engineering, and it would not be easy to scale down the size of this equipment for laboratory work.

VI.   PRACTICAL CONSIDERATIONS

A.   Residence Time

A rapid rate of liquid flow through the separation cell is required to obtain a high through-put for large-scale preparative separations. However, the period during which individual molecules are subjected to the electric field, here called residence time,* must be sufficient to allow optimum focusing to take place.

Furthermore, the flow rate through the apparatus must be within the limits required to maintain good stabilizing properties. For example, very

---

* Also termed flow-through time and transit time.

high flow rates through the Sephadex-packed cell would result in turbulent flow and compacting of the gel bed. With the density gradient and free-flow apparatus turbulent flow is not encountered until very much higher flow rates are reached.

On the other hand, when longer residence time is required for focusing, both the Sephadex and the density gradient methods maintain good stabilizing properties at very slow flow rates. In contrast, the free-flow methods are less suitable for these conditions. With the latter techniques, in the absence of stabilizing media, the velocity of laminar flow must be sufficient to minimize the effects of gravitational and thermal mixing. This aspect is especially important for the vertical orientation of this apparatus.

The time required for focusing will depend on the electrophoretic mobility of the ampholyte, especially the mobility in regions near its isoelectric point, the field strength, the temperature, the migration distance between the two electrodes, the concentration and pH range of the carrier ampholyte, and the properties of the medium such as viscosity or molecular sieving effects. The Ampholine carrier ampholytes have a high charge density in proportion to their molecular size, giving them faster mobilities than proteins. The pH gradient is therefore established first, followed by the slower migration of protein to the focused position. Minimal focusing time will obviously vary with different proteins, depending on their electrophoretic mobilities.

Focusing time will also depend on the range of the pH gradient. With shallow pH gradients spanning only 1–2 pH units the proteins in regions of pH near their isoelectric point will carry only a small charge and will have much slower rates of electrophoretic migration. In a shallow pH gradient they will, therefore, require a longer residence time to obtain a focused condition.

Using small sucrose density gradient columns of his own design, Godson (1970) measured the time required to establish the pH gradient and the focusing time of hemoglobin in Ampholine solutions (pH 3–10). With a field strength of 15 V/cm he obtained a linear pH gradient after 55 min when using 1% Ampholine at room temperature. Under these conditions hemoglobin focused in 2.25 hr whereas at 4°C it took 5 hr to focus.

Godson also showed that a reduction of the Ampholine concentration considerably shortened the focusing time. This effect was confirmed by Catsimpoolas et al. (1974) who measured the minimal focusing time for histidyl tyrosine and soybean trypsin inhibitor in polyacrylamide gels (Chapter 9).

The time course for establishing a pH gradient in the free-flow apparatus (Fig. 12) has recently been given by Just et al. (1975a). The high field strength (90 V/cm) used with this thin-film technique allows a linear pH gradient to be formed across the 11-cm migration path in 15 min. This is approximately

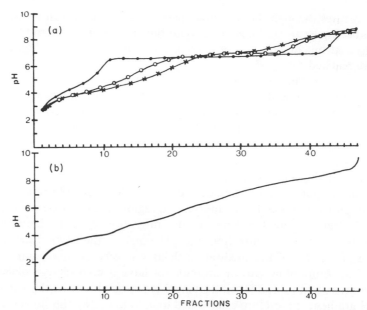

Fig. 12. The pH gradient formed by the free-flow apparatus using 1% Ampholine (pH 3–10). (a) With residence time 6 min and field strength of 45 V/cm (•), 91 V/cm (○), and 125 V/cm (×). (b) Completely established pH gradient, residence time 15 min, field strength 90 V/cm. [From Just *et al.* (1975a).]

the same time required for the establishment of the pH gradient in thin films of polyacrylamide gel with similar applied potential and migration distance (Söderholm and Wadström, 1975).

In 1% Ampholine (pH 3–10) it takes two to four times longer to focus most proteins than to establish the pH gradient. For the continuous-flow isoelectric focusing of proteins, Seiler *et al.* (1970a) used a field strength of 90 V/cm and a residence time of 100 min. On the other hand, Prusík (1974), with an apparatus approximately $4\frac{1}{2}$ times as wide, had a residence time of 150 min with a field strength of 60 V/cm but used only 0.3% Ampholine. A much longer residence time would be required for 1% Ampholine—300 min—if the results obtained from sucrose density gradient experiments by Godson (1970) were applicable to free solution.

A residence time of 7 min for the separation of red blood cells was used by Just *et al.* (1975a). In Ampholine pH 3–10 the cells focused at the same isoelectric points when the position of the sample injection was changed, demonstrating that 7 min was sufficient for focusing conditions. When the same mixture was fractionated in a shallow pH gradient, Ampholine pH 5–7 (Fig. 4b), the rabbit and mouse cells focused at a more acid pH than when the wide range pH gradient was used. This result suggests that the residence time for the shallow gradient was insufficient to allow the cells to reach a true isoelectric focusing position.

Simultaneous application of the sample at two or more inputs, distributed so that components approach the focused isoelectric position from both directions (anodic and cathodic), is by far the most satisfactory test for steady-state focusing conditions (cf. Figs. 7 and 9).

Very much longer residence times were used with the Sephadex and the flowing density gradient apparatus reported by Fawcett (1973) but maximum flow rates were not determined in these experiments. The work was designed to evaluate the long-term stability of the focused position in this continuous-flow equipment and to make a direct comparison between the resolution obtained by these methods and that obtained with the density gradient column.

Shorter residence times have been obtained by a reduction of the migration distance in the flowing density gradient method. For example, with a migration distance of 1 cm and a final field strength of 40–80 V/cm, residence times of 8–15 min are suitable (Section V.C.2). Some reduction of the migration distance in the capillary-stabilized continuous-flow method is possible, but is limited to a distance of 4–5 cm by technical difficulties in collecting multiple fractions and by the irregularities of the flow through the packed bed.

B. Uneven Heat Formation

The electrical resistance increases during the course of isoelectric focusing since the various amphoteric components are distributed according to their isoelectric points. At constant voltage the current will thus gradually fall and consequently less heat is generated in the system. It is normal practice, therefore, with electrofocusing in density gradients and in gels, to increase the applied voltage during the experiment to obtain the highest possible field strength within the limits of the cooling capacity of the apparatus.

With continuous-flow isoelectric focusing, if a constant voltage is applied across the entire length of the separation cell, more heat will be formed at the inflow end where the mixed ampholytes have a higher conductance. The difference in conductance of the ampholytes in the mixed state and in the focused state will depend on the pH range and the nature of the ampholytes. The difference is greatest for the wide range pH gradients and can result in a 10- to 15-fold difference with Ampholines pH 3–10. Thus, when using the wide range Ampholine, if an equal voltage is applied across both ends of the cell, then the heat formed at the inflow end will be approximately 10–15 times that at the outflow. This situation is unsatisfactory since the limitation on voltage to prevent excessive heating at the inflow end results in the remainder of the cell operating well below its cooling capacity. Furthermore, to obtain better resolution, it is advantageous to subject the ampholytes to the maximum voltage gradient just prior to leaving the apparatus, since the width of the focused zone is inversely proportional to the square root of the field strength (Svensson, 1961).

The problems caused by the higher conductance of the unfocused carrier ampholytes may be solved in a number of ways:

1.   The carrier ampholytes are fractionated first in a prerun through the apparatus. Fractionated ampholytes, in the appropriate order, are fed into the separation cell, thus directly establishing the pH gradient. This is the method used by Prusík (1974) (Section V.A.2). Note, only moderate applied potentials are required to fractionate the carrier ampholytes and the uneven heat distribution in the prerun presents no difficulties.

2.   By construction of a separation cell with increased cooling efficiency at the inflow end. For example, the cell could be wedge shaped in cross section with smaller thickness at the inflow end to allow more efficient dissipation of heat.

3.   By precooling the inflow solution below the cell temperature to allow for a temperature rise. This method is only suitable when differences in conductance are small (narrow pH range ampholytes) and when rise in temperature does not cause disturbing thermal convections.

4.   Electrode compartments divided into a number of sections, each with separate electrode. Maximum voltage is applied across the outflow end and, by means of external resistances in the circuit, the voltage applied across the other sections is reduced in a stepwise manner (Section V.C.2).

5.   Advantage may be taken of the resistance of the electrode solutions. The electrodes are positioned to give maximum voltage across the outflow end and the current to the remainder of the cell is supplied by the conducting layers of the electrode solutions. In this way the voltage applied across the cell is steadily reduced from outflow to inflow ends (Section V.B).

In addition to differences in heat formation at the inflow and outflow ends the heating is also nonuniformly distributed across the cell. Initially, the electrical conductance of the mixed ampholytes is uniform across the cell but as the pH gradient develops the conductance will vary at each position. The first regions of the pH gradient to develop are near the anode and cathode (Fig. 12) and, because of the higher resistance of these focused ampholytes, warmer zones begin to develop in these regions. Davies (1975), using an infrared Thermovision camera, has investigated the heat formation during isoelectric focusing in thin slabs of polyacrylamide gel. At the beginning of an experiment using pH 3–10 Ampholine he found an increase in the temperature of the gel near the electrodes. The gel was warmer near the anode than near the cathode. During the formation of the pH gradient these warmer zones broaden and move toward the center of the gel until finally the temperature is at a maximum in the center and decreases toward each electrode. Similar temperature gradients are expected in other stabilizing media.

The warmer zone in the center of the pH gradient when focusing is established is due to the lower conductance of ampholytes in this region. In principle, by careful selection of the composition of the carrier ampholytes and the adjustment of the concentration of the individual components, uniform conductance can be obtained along most of the pH gradient, but because of concentration of hydrogen and hydroxyl ions, higher conductance is always encountered close to the electrodes. Variations in conductance as the pH gradient develops are unavoidable.

With the Sephadex and density gradient continuous-flow apparatus, and with the horizontal free-flow apparatus, the stabilization against convection is usually sufficient to accommodate these small variations in temperature along the pH gradient. On the other hand, with the vertical form of free-flow apparatus, thermal convection is unrestricted and any localized heating especially at the outflow end will result in a loss of resolution.

C.   pH Gradient Stability

A stable pH gradient must be maintained for the focused zone to emerge from the same outflow position during prolonged periods of continuous-flow isoelectric focusing. Properties most likely to disrupt the stability of the pH gradient are precipitation, changes in the pattern of liquid flow caused by settling or channeling of the cell packing, or changes in electroosmotic flow in the cell. Formation of precipitates during some electrofocusing separations is the most serious of these defects and is discussed in more detail in the next section.

Electroosmotic flow, the bulk movement of liquid toward one of the electrodes, is caused by the presence, within the electric field, of surface charges on the solid supporting medium or on the walls of the separation cell. The movement of liquid is greatest in capillary systems and with solutions of low ionic concentration. Since the surface charge is pH dependent, electroosmosis during isoelectric focusing will result in complex flow movements.

Bulk transfer of liquid has been observed during isoelectric focusing in Sephadex (Fawcett, 1975) but the movement in Sephadex G-75 and G-100 is not sufficient to disrupt the pH gradient significantly. Far greater liquid movement is observed with the less porous grades of Sephadex, presumably because of higher proportion of charged groups, and these grades are unsuitable for use in isoelectric focusing (Section V.B).

With the flowing density gradient and free-flow methods electroosmotic flow can result only from charge groups on the surface of the cell walls. It is only in the free-flow apparatus where the gap between the two surfaces is small that the effect is of any importance. Because electroosmotic flow takes place adjacent to the cell walls it results in a counterflow of liquid in the opposite direction along the center of the cavity (assuming negligible electro-osmotic flow through the membranes at the edges of the cell). Liquid movement of this type will impart a parabolic profile to the shape of any zone when viewed in a plane perpendicular to the bulk liquid flow through the cell. This phenomenon has been observed experimentally in the free-flow electrophoresis of particles (Strickler and Sacks, 1973) where it results in a decrease in resolution between closely separated zones. Similar effects are likely during free-flow isoelectric focusing but, as previously stated, the rate of electroosmotic flow is pH dependent and will vary across the cell.

For isoelectric focusing in a narrow pH range it should be possible, by suitable choice of material for the cooling surfaces, to construct separation cells with negligible electroosmotic properties over the required pH range, but during continuous operation it is important that the surface does not become charged due to the effect of adsorption or precipitation.

It is well established that the pH gradient slowly drifts toward the cathode if isoelectric focusing in gels or density gradient columns is continued for long periods. The drift is most pronounced at the alkaline end of the pH gradient and is more rapid in polyacrylamdie gel or Sephadex gel than in the sucrose density gradient. It has been suggested (Fawcett, 1975) that the drift is caused by a lack of equilibrium between the electrophoretic migration of an ampholyte to the isoelectric point and the transport by diffusion away from the isoelectric point.

Results of this pH drift have been observed in continuous-flow experiments (Fawcett, 1973). For example, using a continuously flowing density gradient with 54 outflow tubes, the position at which the hemoglobin is collected is displaced three tubes toward the cathode when the residence time is doubled. A similar displacement toward the cathode is observed when the applied voltage is doubled.

However, it must be emphasized that these are extreme changes of conditions and that the small fluctuations in residence time (i.e., small fluctuations in flow rate) or field strength that are normally encountered in continuous-flow experiments have only negligible effect on the pH gradient stability.

D. Precipitation

A serious limitation of almost all isoelectric focusing techniques is that some proteins have a low solubility at their isoelectric point. This often causes precipitation of the focused protein and, with the LKB density gradient column for example, these precipitates often sediment into more dense regions and interfere with the detection and isolation of other protein zones.

Precipitation in continuous-flow methods is also a major problem, particularly with the apparatus stabilized with Sephadex or granulated polyacrylamide gel. Very finely divided precipitates in the form of turbid zones do not interfere with the liquid movement and will flow through the packed gel bed. It is when aggregation of the precipitates occurs and the large particles stick to the gel surface that the effects of "clogging" become apparent. The positions of focused zones emerging from the apparatus are no longer constant, due mainly to the distorted liquid flow through the gel bed but also possibly as the result of electroosmosis caused by regions where the gel particles have acquired a surface charge from the precipitated protein.

Both the free-flow apparatus and the flowing density gradient apparatus are capable of operating with a somewhat higher degree of precipitation. In the flowing density gradient fine precipitates may sediment to lower (more dense) regions and contaminate other focused zones. It is only when the precipitates are gelatinous and attach to the walls of the apparatus and clog the tubing to the pump unit that continuous operation becomes impossible.

With the free-flow apparatus positioned in the horizontal plane, precipitates will sediment onto the surface of the lower plate and seriously disrupt continuous operation. Moderate precipitation in the vertical form of the free-flow apparatus is not so disastrous. In this apparatus, if the precipitate from a focused component begins to sediment, it will move with the liquid flow; focused zones also travel with the liquid. Thus the precipitate and the soluble portion are collected together and in this apparatus the precipitates do not necessarily interfere with neighboring focused zones. As with the density gradient apparatus continuous operation is impossible if heavy precipitation occurs and material becomes attached to the walls of the separation unit.

The addition of nonionic detergents (Godson, 1970; Friesen *et al.*, 1971) often reduces precipitation, or where heavy precipitates normally occur the detergents keep the particles in a finely suspended form.

Allen and Humphries (1975) have reported the use of the zwitterionic detergents Empigen BB and Sulfobetaine DLH for the isoelectric focusing of biological membranes. They found that these compounds in low concentrations [0.1–0.5% (w/v)] were very effective in solubilizing membranes from red blood cells and milk fat globules during isoelectric focusing in sucrose density gradients. Since these zwitterionic detergents have strong acid and basic groups they are isoelectric over the pH region 3.5–9.0 and do not interfere with the formation of the pH gradient. These detergents should prove very useful in reducing precipitation in continuous-flow experiments. It must be emphasized, however, that some enzymes are inactivated by these materials (Allen and Humphries, 1975).

The aggregation of cells and cell particles during isoelectric focusing also presents major difficulties. The aggregation of red blood cells is prevented by prior washing with isotonic saline containing EDTA, but Just *et al.* (1975b) found that this treatment did not prevent aggregation of the light mitochondrial fraction of rat liver. With the latter material it was found that aggregation during isoelectric focusing could be prevented by a number of polyanionic substances such as heparin, chondroitin sulfate, polyvinyl sulfate, and dextran sulfate (Just *et al.*, 1974, 1975b).

E. Sample Capacity

For continuous-flow methods the total sample load per day is proportional to sample concentration and to the flow rate. However, the maximum load is very dependent on the solubility of the components of the mixture and especially the solubility in the regions of their isoelectric points if precipitation is to be avoided.

Where solubility is not the limiting factor consideration must be given to density and viscosity effects. During isoelectric focusing ampholytes are focused into concentrated zones to give regions of increased density and increased viscosity which impose practical limits on the maximum load that can be applied.

Of the stabilizing systems available, the one best suited to stabilize zones of high concentration and density is the Sephadex-packed bed where the capillaries between the particles completely prevent gravitational mixing. Unfortunately, the viscosity increase associated with the very high sample concentration seriously distorts the liquid flow in the packed bed. The increased viscosity of the focused zone reduces the flow of the zone through the bed. With continuous sample application more protein will be focused in this slow moving zone, thus further increasing the viscosity. One can postulate a spiraling situation developing. In practice this does not seem to happen but the resolution of components is seriously affected when very high sample loads are used.

In density gradient stabilization, as the components begin to focus, the local increase in density results in a widening of the zone by spreading downward under gravity, until a positive density gradient is reestablished. The density gradient stabilization imposes a strong viscosity gradient and any viscosity increase in the focused zone is of little consequence. A constant horizontal liquid flow is always maintained at all levels by the hydrodynamic effect of the density gradient.

In the horizontal free-flow apparatus the gravity-induced spreading of focused zones is very slow because the thin horizontal liquid film (<1 mm) exerts only a small hydrostatic head. But the spreading is time dependent, and the effect is more important with the long residence times (slow flow rates) often required for isoelectric focusing and when high loading is attempted. Any increase in viscosity of the focused zone results in a slower movement through the cell, which is compensated by the gravitational effect if the cell is positioned with a slight downward inclination.

The gravitational effect on a concentrated zone in the vertical form of the free-flow apparatus causes a downward movement faster than the main liquid flow. This effect is opposed by the viscosity of the concentrated zone which reduces the flow of the zone through the cell. Unfortunately, with the slow flow rates required for isoelectric focusing the effect of increased density predominates and sets an upper limit to the sample load that can be applied to this technique.

Until more data become available it is difficult to compare the total sample capacity of the different forms of continuous-flow equipment used for isoelectric focusing. Some apparatus have been deliberately constructed with a small separation chamber and some are operated at slow flow rates and well below the maximum capacity in order to economize in the consumption of the costly carrier ampholytes during test experiments. The effective volumes of the separation chambers and the reported residence times are listed in Table I. Calculated total volume output, though not necessarily the maximum values obtainable, give some indication of the potential capacity of this equipment.

TABLE I

Capacity of Continuous-Flow Isoelectric Focusing Equipment

Type	Cell volume (ml)	Migration path length (cm)	Residence time[a] (min)	Flow rate (ml/hr)	Ref.
Free flow					
Vertical (proteins)	20	11	50–100	10–20	Seiler et al. (1970a)
Vertical (cells)	20	11	7[b]	60	Just et al. (1975a)
Horizontal	110	50	150	44	Prusík (1974)
Capillary stabilized	200	23	1000	11	Fawcett (1973)
	140	14	420	20	Gianazza et al. (1975)
	345	20	170	125	Hedenskog (1975)
Density gradient					
Long migration path	240	23	600	24	Fawcett (1973)
Short migration path	140	1	8–15	560–1050	Fawcett (unpublished)

[a] The time quoted is not necessarily the minimum time for focusing.

[b] Cells are injected at a position halfway down the cell. Residence time for carrier ampholytes is 15 min.

F. Carrier Ampholytes

At present the development of large-scale continuous-flow isoelectric focusing is undoubtedly limited by the cost of the carrier ampholytes. It is only when valuable material is fractionated, and when isoelectric focusing offers clear advantages over all other separation methods, that it is economical to use this process. However, inspection of price lists of biochemical products reveals that many substances listed are sufficiently valuable to justify purification by isoelectric focusing techniques. An important consideration is the excellent yield usually obtained with isoelectric focusing methods—an essential requirement when fractionating valuable materials.

A large part of the cost in manufacturing these ampholytes is certain to come from the extensive processing, both before and after synthesis, which is necessary to give products suitable for laboratory use. Solutions of the carrier ampholyte must be relatively transparent in the 260–280 nm region to allow for the detection of proteins by ultraviolet light absorption. Furthermore, carrier ampholytes used for gel isoelectric focusing must be free of contaminating material likely to produce permanently stained zones from established procedures. Both these requirements necessitate elaborate purification of acids and bases used in the synthesis and detailed control of conditions during the condensation process. In addition, the reaction product is subjected to electrophoretic purification and is suitably blended to obtain products with good conductivity over selected pH ranges. All these processes add extensively to the cost of this material.

It would seem that less refined carrier ampholytes would be adequate for large-scale preparative isoelectric focusing. Ampholytes possessing considerable ultraviolet absorption would be quite suitable since in large-scale work proteins with known isoelectric points are identified by pH measurements and the analysis of fractions by light absorption is unnecessary. It would appear that products obtained from the reaction of certain polyethylene amines with acrylic acid, without electrophoretic fractionation, are suitable for preparative isoelectric focusing (Vinogradov et al., 1973). If such products form nonlinear pH gradients, they are still acceptable provided electrical conductance is maintained throughout the pH range. Less refined carrier ampholytes of this nature are obviously very much cheaper to produce since chemicals for this synthesis are not particularly expensive. If carrier ampholytes are prepared from less pure materials, it is important that they are still readily separated from fractionated material. Therefore, there should be no increase in the molecular size of the ampholytes.

Clearly the cost of separations by isoelectric focusing is reduced if satisfactory results are obtained using lower concentrations of the carrier ampholytes. In addition, the problem of heat removal is less since a lower ampholyte concentration reduces the electric current and less Joule heat is produced. Moreover, focusing is more rapid because of the increased electrophoretic mobility at the lower ampholyte concentrations, but in very low ampholyte concentrations regions of extremely low conductance can develop, resulting in hot zones and uneven field strength. Furthermore, with high sample loads the separation of closely focused components will deteriorate as the carrier ampholyte concentration is reduced. Two similar components of a mixture will separate only if carrier ampholytes are present in sufficient concentration to prevent abrupt changes in the pH gradient and to maintain a reasonably uniform electrical conductance. Thus, for a given degree of resolution, there are optimal values for the sample load and the carrier ampholyte concentration for each individual ampholyte mixture, but no systematic investigation of this topic appears to have been made.

An important aspect in any consideration of the economics of large-scale isoelectric focusing is the feasibility of the recovery and recycling of carrier ampholytes. In the absence of sucrose or other density-stabilizing material moderate recovery could be obtained using ultrafiltration methods. Separation by alcohol or acetone precipitation methods is a further possibility when stable porteins are involved. In any case it would seem inevitable that the recovered carrier ampholytes would become contaminated with any low-molecular-weight ampholytes present in the sample mixture. Recovery of ampholytes from sucrose density gradient isoelectric focusing presents a greater problem since the viscosity of the solution interferes with the ultra-filtration process.

The use of partial hydrolysates of proteins as substitutes for the commercial carrier ampholytes is another possible way of reducing the cost of isoelectric focusing. Originally investigated by Svensson (1962) these materials have been tested more recently by Blanický and Pihar (1972) who investigated their use with both the density gradient and polyacrylamide gel techniques. They are available commercially as bactopeptones and other culture media and are relatively cheap. Before use as ampholytes in isoelectric focusing it is recommended that the less soluble peptides be removed by precipitation with aqueous ethanol. However, we find that this procedure does not remove all the peptides which are insoluble at their isoelectric point, and frequently precipitated or turbid zones are obtained during iso-electric focusing with this material. Naturally, solutions of these peptides have considerable absorption at 280 nm but as previously stated this presents no disadvantage with large-scale preparative work where fractions can be identified by pH measurement.

During isoelectric focusing with peptide mixtures prepared in this manner, regions of low conductance often develop because of the lack of suitable isoelectric components in some parts of the pH gradient.

Finally, in search of cheaper alternatives to carrier ampholytes, the early work with artificial pH gradients obtained by diffusion of buffer solutions (Kolin, 1954) should be reinvestigated. Gradients prepared in this way may find applications with continuous-flow isoelectric focusing techniques with very short separation distances (Section V.C.2). In fact, the continuous-flow approach to this method of gradient formation solves some of the earlier difficulties. These gradients, of course, no longer possess the stability of the natural pH gradient (those obtained by ampholytes) and will require carefully controlled conditions to obtain reproducible separations. Large-scale equipment fitted with continuous monitoring and automatic feedback control of electric potential and flow rate should be capable of maintaining such conditions.

Recently Troitskii *et al.* (1974) have described an entirely new technique for creating stable pH gradients from ordinary buffers. This is achieved by varying the dielectric property of the medium with gradients of organic solvents such as ethanol, dioxane, or glycerol. For example, borate buffers and glycerol (which forms complexes with borate) provide a pH gradient, stable for 12 days, with a range of 4.5 pH units. If satisfactory as a general method of forming pH gradients, these techniques will encourage the use of continuous-flow isoelectric focusing on a very large scale.

VII. CONCLUSIONS

A number of experiments have demonstrated that continuous-flow iso-electric focusing is capable of high resolution for the separation of test mixtures and is able to maintain steady-state conditions over long periods, but few applications to general practical problems have been reported.

Until more experimental results are available it is not possible to give detailed comparisons between the various stabilizing methods or the different apparatus designed for this technique, but some general conclusions can be drawn.

1. The capillary-stabilized method (Sephadex) and the horizontal free-flow method are limited to conditions where no precipitation occurs.

2. Moderate precipitation during isoelectric focusing causes least disturbances to separated zones in the vertical free-flow apparatus.

3. When components precipitate and form aggregates or when cells and cell particles clump and form aggregates none of the existing methods is satisfactory.

4. The vertical free-flow apparatus allows rapid isoelectric focusing of cells and cell particles and appears more suitable than the flowing density gradient for this type of separation.

5. Although the flowing density gradient is very satisfactory for the separation of proteins, the recovery of material from the viscous media is more difficult.

6. Methods using short migration distances are economical on power consumption and allow designs for large capacity, but place high demands on stabilization and parallel flow and require efficient removal of the very thin zones.

7. For the separation of proteins on a relatively small scale (1–2 g) batch methods such as Sephadex slab (Chapter 6) or large-capacity short density gradient columns (Chapter 2) are more suitable.

Although more operational data are necessary before the potential of continuous-flow isoelectric focusing can be assessed accurately, the results obtained should stimulate further developments in this technique. Certainly, the present cost of the process prohibits its application to general preparative methods but it has attractive possibilities in large-scale work for the more special separation problems requiring high resolving power.

ACKNOWLEDGMENTS

I wish to thank Dr. G. Werner for supplying a photograph of his apparatus (Fig. 2b) and details of his results in advance of publication. Permission for reproduction of published figures is gratefully acknowledged.

REFERENCES

Allen, J. C., and Humphries, C. (1975). In "Isoelectric Focusing" (J. P. Arbuthnott and J. A. Beeley, eds.), pp. 347–353. Butterworths, London.
Barrollier, V. J., Watzke, E., and Gibian, H. (1958). Z. Naturforsch. 13b, 754–755.
Blanický, P., and Pihar, O. (1972). Collect. Czech. Chem. Commun. 37, 319–325.
Brattsten, I. (1955). Ark. Kemi 8, 205–226.
Brattsten, I., and Nilsson, A. (1951). Ark. Kemi 3, 337–345.
Catsimpoolas, N., Campbell, B. E., and Griffith, A. L. (1974). Biochem. Biophys. Acta 351, 196–204.
Davies, H. (1975). In "Isoelectric Focusing" (J. P. Arbuthnott and J. A. Beeley, eds.), pp. 97–113. Butterworths, London.
Durrum, E. L. (1951). J. Amer. Chem. Soc. 73, 4875–4880.
Durrum, E. L., and Pickels, E. G. (1958). Can. Patent 644,875, issued 1962.
Fawcett, J. S. (1973). Ann. N.Y. Acad. Sci. 209, 112–126.
Fawcett, J. S. (1975). In "Progress in Isoelectric Focusing and Isotachophoresis" (P. G. Righetti, ed.). pp. 25–37. North-Holland, Amsterdam.
Friesen, A. D., Jamieson, J. C., and Ashton, F. E. (1971). Anal. Biochem. 41, 149–157.
Gianazza, E., Pagani, M., Luzzana, M., and Righetti, P. G. (1975). J. Chromatogr. 109, 357–364.
Godson, G. N. (1970). Anal. Biochem. 35, 66–76.
Grassmann, W., and Hannig, K. (1950a). Angew. Chem. 62, 170.
Grassmann, W., and Hannig, K. (1950b). Naturwissenschaften 37, 397.
Hannig, K. (1961). Z. Anal. Chem. 181, 244–254.
Hannig, K. (1969). In "Modern Separation Methods of Macromolecules and Particles" (T. Gerritsen, ed.), pp. 45–69. Wiley (Interscience), New York.
Hannig, K. (1972). Tech. Biochem. Biophys. Morphol. 1, 191–232.

Hedenskog, G. (1975). *J. Chromatogr.* **107**, 91–98.

Just, W. W., León, J., Werner, G., Thobe, J., and Seiler, N. (1973). *Abstr. Int. Congr. Biochem.*, *9th, Stockholm* p. 30.

Just, W. W., León, J. O., and Werner, G. (1974). *Abstr. FEBS Meeting, 9th, Budapest* p. 430.

Just, W. W., León, J. O., and Werner, G. (1975a). *Anal. Biochem.* **67**, 590–601.

Just, W. W., León, J. O., and Werner, G. (1975b). *In* "Progress in Isoelectric Focusing and Isotachophoresis" (P. G. Righetti, ed.). pp. 265–280. North-Holland, Amsterdam.

Kolin, A. (1954). *J. Chem. Phys.* **22**, 1628–1629.

Mel, H. C. (1960). *Science* **132**, 1255–1256.

Mel, H. C. (1964). *J. Theor. Biol.* **6**, 307–324.

Philpot, J. St. L. (1940). *Trans. Faraday Soc.* **36**, 38–46.

Philpot, J. St. L. (1965). Brit. Patent. 1,150,722, issued 1969.

Philpot, J. St. L. (1973). *In* "Methodological Developments in Biochemistry" (E. Reid, ed.), **2**, pp. 81–85. Longman, London.

Prusík, Z. (1974). *J. Chromatogr.* **91**, 867–872.

Radola, B. J. (1973a). *Abstr. FEBS Special Meeting, Dublin* No. 64.

Radola, B. J. (1973b). *Biochem. Biophys. Acta* **295**, 412–428.

Rilbe, H. (1970). *Proc. Colloq., Protides Biolog. Fluids, 17th, Bruges, 1969* **17**, 369–382.

Rilbe, H. (1973). *Ann. N.Y. Acad. Sci.* **209**, 80–93.

Robinson, L. C. (1973). *In* "Methodological Developments in Biochemistry" (E. Reid, ed.), **2**, pp. 87–94. Longman, London.

Seiler, N., Thobe, J., and Werner, G. (1970a). *Hoppe-Seyler's Z. Physiol. Chem.* **351**, 865–868.

Seiler, N., Thobe, J., and Werner, G. (1970b). *Z. Anal. Chem.* **252**, 179–182.

Söderholm, J., and Wadström, T. (1975). *In* "Isoelectric Focusing" (J. P. Arbuthnott and J. A. Beeley, eds.), pp. 132–142. Butterworths, London.

Strain, H. H. and Sullivan, J. C. (1951). *Anal. Chem.* **23**, 816–823.

Strickler, A., and Sacks, T. (1973). *Ann. N.Y. Acad. Sci.* **209**, 497–514.

Svensson, H. (1961). *Acta Chem. Scand.* **15**, 325–341.

Svensson, H. (1962). *Arch. Biochem. Biophys. Suppl. 1* 132–138.

Svensson, H., and Brattsten, I. (1949). *Ark. Kemi* **1**, 401–411.

Svensson, H., and Valmet, E. (1955). *Sci. Tools* **2**, 11–13.

Thomson, A. R., Murdoch, R., Gibson, J. A., Smyth, M. J., Atherton, R. S., and Laws, J. F. (1973). *Abstr. FEBS Special Meeting, Dublin* No. 63.

Tippetts, R. D. (1965). Ph.D. Thesis, Univ. of California, Berkeley.

Troitskii, G. V., Zaryalov, V. P., and Abramov, U. M. (1974). *Dokl. Akad. Nauk USSR* **214**, 955–958.

Vinogradov, S. N., Lowenkron, S., Andonian, M. R., Bagshaw, J., Felgenhauer, K., and Pak, S. J. (1973). *Biochem. Biophys. Res. Commun.* **54**, 501–506.

Williams, R. R., and Waterman, R. E. (1929). *Proc. Soc. Exp. Biol. Med.* **27**, 56–59.

Winsten, S., Friedman, H., and Schwartz, E. E. (1963). *Anal. Biochem.* **6**, 404–414.

Valmet, E. (1969). *Sci. Tools* **16**, 8–13.

8 ISOELECTRIC FOCUSING IN FREE SOLUTION

Johan Bours

Klinisches Institut für Experimentelle Ophthalmologie
Universität Bonn
Bonn-Venusberg, West Germany

I. INTRODUCTION

In this chapter a number of different methods of isoelectric focusing in free solution are compared and evaluated according to their specific merits and practical applicability. In addition, various results of each method covering representative experiments are compared and presented in table form.

II. COMPARISON OF DIFFERENT METHODS OF ISOELECTRIC FOCUSING IN FREE SOLUTION

When performing isoelectric focusing in free solution, the liquid system must be stabilized against a disturbing convection. In each of the methods listed in Table I, the requirements for stabilization are met.

TABLE I   Comparison of Different Methods of Isoelectric Focusing Experiments in Free Solution

No.	Method of isoelectric focusing	Method for stabilization against convection	Substances present (except sample)	Scale	Amount of sample	Period	pH determined	Presence of protein precipitates	$\partial pH/\partial x$ for 1 pH unit/cm	Ref.
1	Sucrose density gradient isoelectric focusing	Sucrose density gradient	Ampholine carrier ampholytes; sucrose	Semi-preparative; preparative	5–30 mg	2–3 days	In eluted fractions	Disturbant	0.43	Vesterberg and Svensson (1966), Haglund (1967, 1971), Bours (1973d)
2	Zone convection isoelectric focusing	Vertical density gradient due to "thermal diffusion"	Ampholine carrier ampholytes	Preparative	Up to 1 g	2(–5) days	Directly in the apparatus	No influence	0.06	Valmet (1969), Haglund (1971)
3	Free zone isoelectric focusing	Slow rotation of the electrophoresis tube	Ampholine carrier ampholytes	Analytical	~100 µg	30 hr	In fractions removed from the electrophoresis tube	Disturbant	0.25	Lundahl and Hjertén (1973)
4	Free isoelectric focusing in coils of polyethylene tubing	Mechanically	Ampholine carrier ampholytes, sucrose	Semi-preparative	5–50 mg	24–30 hr	In fractions cut from the coil	No influence	0.04	Macko and Stegemann (1970), Bours (1972, 1973c)
			Ampholine carrier ampholytes without sucrose						0.04 0.04 (mean)	Bours (1975), Domagk et al. (1973), Zech and Zürcher (1973)
5	Analytical isotachophoresis (LKB-Tachophor 2127)	Mechanical	Ampholine carrier ampholytes as spacer ions; buffer ions	Analytical	5–10 µg	12–20 min	Does not apply	No precipitation occurs		Kjellan et al. (1975), see LKB application notes

1.   In density gradient isoelectric focusing, stabilization is obtained by sucrose. This requires subsequent dialysis of separate liquid fractions to obtain a protein preparation. This method, which was developed in an applicable form by Vesterberg and Svensson (1966), has been applied widely as a (semi-) preparative method with considerable success. Nevertheless, the density gradient technique is a very reliable and exceptional method for determining isoelectric points of major components of a complex protein mixture at a constant temperature of 4°C (Bours, 1973d).

2.   Valmet's (1969, 1970a) zone convection electrofocusing method requires no stabilizing medium as such, and the apparatus can be loaded with a considerable amount of sample, e.g., proteins.

3.   The free zone electrophoresis method of Hjertén (1967a,b), which is only applicable on an analytical scale, has a good reproducibility, is totally free of electroendosmosis, and is effectively stabilized against disturbing convective deformation by rotation of the electrophoresis tube. The free zone isoelectric focusing method is a modification of Hjertén's free zone rotating tube technique because of the incorporation of the isoelectric focusing principle.

4.   Free isoelectric focusing is conducted in coils of polythylene tubing, and stabilization is achieved mechanically. This method, which was originally designed by Macko and Stegemann (1970), is preeminently suitable for the purification of enzymes and proteins on a semipreparative scale.

III.   ZONE CONVECTION ISOELECTRIC FOCUSING

The preparative method of zone convection isofocusing was designed by Valmet (1969, 1970a) for proteins according to the modified principle based on a membrane electrolyzer which was used by several other workers and had been reviewed earlier by Svensson (1948). A schematic drawing of the apparatus is shown in Fig. 1. The apparatus consists of two corrugated separate parts, the trough and the lid, forming a series of "U-tube units." The current thus forms a natural density gradient in each U-tube due to thermal diffusion concomitant with isofocusing, and the proteins are collected at the bottom of each trough (Fig. 1b). An extensive description of the apparatus and the operation of this method is given by Haglund (1971). Also, a short comprehensive description is given by Catsimpoolas (1973). Although basically sound in principle, reproducibly satisfactory results cannot be obtained by this method. This is because there is the possibility of remixing the separated fractions when the lid is lifted from the apparatus after separation. Thus, any resolution obtained may be partially lost.

However, it is still difficult to judge the resolving power and the separational ability of the zone convection isofocusing method, because Valmet

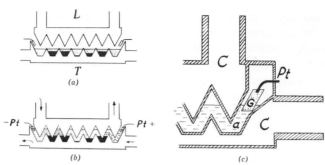

Fig. 1. Zone convection isoelectric focusing. Schematic drawing of the zone convection electrofocusing apparatus. (a) The filled trough T, before or after the experiment. The separated protein zones are illustrated by dark areas. The lid L, lifted up as before or after the run. (b) Both parts put together, as during the experiment. Notice the different positions of the ampholyte solution level. Both parts are water-cooled. The arrows show the circulation of cooling water; Pt indicates the electrodes. (c) Details of the electrode: Pt, platinum wire; G, sintered glass filter; a, ampholyte solution; C, cooling water. [From Haglund (1971) (originally Valmet). Reproduced from *Methods of Biochemical Analysis* by permission of John Wiley & Sons, Inc.]

(1968, 1969, 1970a,b) failed to publish the precise analysis of the distribution of the proteins formed in each trough, which can be easily appraised by any electrophoretic method. Because this apparatus is not commercially available and must be specially constructed in the laboratory, currently only a few authors have succeeded in using this method (Kiryukhin, 1972; Tolkacho, 1974; Troitski *et al.*, 1974a,b). However, a modification of Valmet's method was used with reasonable success in isolating a hormone preparation from a pituitary extract (Denckla, 1974). An improvement in zone convection iso-focusing was made by Kiryukhin (1972) who, upon reevaluating Valmet's method, performed electrofocusing in a similar horizontal apparatus with extended lengths, thus increasing the number of U-tube units from 30 to 48. Troitski *et al.* (1974a) and Tolkacho (1974) succeeded in isolating pure albumin fractions from rabbit serum at pH 4–6, using Kiryukhin's improved method (1972).

Bodwell and Creed (1973, cited by Fawcett, 1975), using a Valmet type of apparatus, seemed to obtain good results in separating components from a number of protein preparations. Fawcett (1975) constructed a similar apparatus for zone convection isofocusing; the apparatus consists of a long narrow Perspex trough with 38 compartments. He reported good results when proteins with low isoelectric points were isofocused, such as ovalbumin, but failed to obtain sharply focused zones at the alkaline end of the pH gradient with a sample of whale myoglobin.

Although Haglund (1971) and Catsimpoolas (1973) have found reasonably good prospects for the modification of Valmet's method, which is essentially a continuous-flow operation, in my opinion the outlook for the general use of this method appears limited.

However, recently Talbot and Caie (1975) successfully applied Valmet's principle of zone convection isoelectric focusing in a horizontal trough apparatus of their own design. With this apparatus, Talbot (1975) obtained a good separation and reproducible measurements of the isoelectric points of two interconvertible components from foot-and-mouth disease virus.

IV. FREE ZONE ELECTROPHORESIS

In carrier-free zone electrophoresis one of the main problems is how to stabilize separated zones or bands without the use of supporting media. In order to eliminate convection during free electrophoresis, Hjertén (1967a,b) succeeded in designing a carrier-free zone electrophoresis apparatus in which stabilization against convection was achieved by slowly rotating the electrophoresis tube about its longitudinal axis during a run. Figure 2 describes how stabilization of zones was achieved by employing this rotation. The electrode vessels and the horizontal revolving electrophoresis tube are immersed in a water bath of constant temperature, and mounted on a carriage. During an electrophoretic run, this system is moved at preset time intervals, and the migrating substances can be detected by a sensitive stationary ultraviolet scanning system.

Since numerous technical problems have been overcome (Hjertén, 1967a,b), free zone electrophoresis as an analytical tool has extensive applications in separations. Low-molecular-weight substances (e.g., inorganic ions, amino acids, purine and pyrimidine bases, nucleosides, and nucleotides) as

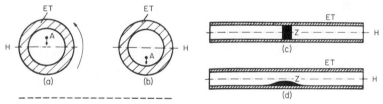

Fig. 2.   Free zone electrophoresis. Drawings illustrating the principle involved in the stabilization of zones by rotation of the horizontal electrophoresis tube ET around its long axis. H represents the horizontal plane. A is a liquid element that has a higher density than the surrounding buffer solution. If we assume that gravity and frictional forces are the only forces acting on the liquid element, the direction of its movement must always be downward. Therefore, the liquid element moves away from the tube wall when it is in the upper half of the cross section (a) and toward the wall when in the lower half (b). According to these simplified diagrams the liquid element is forced by the rotation of the tube to move to-and-fro relative to the tube wall, which is equivalent to zone stabilization. If rotation is stopped, the sharp zone Z in (c) will spread rapidly, as shown in (d). [From Hjertén (1967b). Reproduced by permission of Elsevier Scientific Publishing Company.]

well as larger molecules, particles, and even whole cells can be satisfactorily separated by Hjertén's method.

The reproducibility and minimum zone spreading of this method have proved to be excellent, and electroendosmosis, if present, is very low. To evaluate whether the rotation of the electrophoresis tube effectively stabilizes against disturbing convection, the direction of the current was reversed immediately after the separation, and this reversed electrophoresis again resulted in one peak. In this respect it was shown that the convective deformation of the separated zones is negligible.

In analyzing the complex mixtures of proteins, free zone electrophoresis has been used for serum proteins, amniotic fluid (Fig. 3), and milk proteins (Table II). The free zone electrophoresis method has been successful in testing

TABLE II Free Electrophoresis According to Hjerté

				Determined by free electrophoresis			
No.	Protein(s)	Enzyme commission No.	Quantity analyzed	Maximum running time (min)	No. of peaks observed	Electrophoretic mobility $(cm^2 V^{-1} sec^{-1})$	Purity
1	Normal human serum		10 μl serum diluted 1:2.2	36	8		Heterogeneous
2	Amniotic fluid		4 μl	42	4		Heterogeneous
3	Maternal plasma		7 μl	40	7		Heterogeneous
4	Proteins of cow's milk		10 μl	40	8		Heterogeneous
5	Cellulase from Penicillium notatum	—	10 μl 0.15%	180	1		Homogeneous
6	Proteinase from Arthrobacter sp.	—	10 μl 0.4%	244	1	2.57×10^{-5} at pH 8.6	Homogeneous
7	Pyruvate dehydrogenase	—	20 μl 1%	36	1	—	Homogeneous
8	Neurotoxin from the venom Naja nigricollis	—	10 μl	132	1	5.0×10^{-5} at pH 8.7	Homogeneous
9	Dextranase from Cytophaga sp.	—	—	204	1	2.71×10^{-5} at pH 8.7	Homogeneous
10	Hemagglutination inhibitor from human serum against Picorna viruses	—	—	24	1(2)	—	Nearly homogeneou
11	α-Hydroxysteroid dehydrogenase from Pseudomonas testeroni	E.C.1.1.1.50	—	—	—	—	Homogeneous
12	β-Glucosidase from Trichoderma viride	E.C. 32.1.21	10 μl 0.16%	110	1	1.49×10^{-5} at pH 8.0	Homogeneous
13	5'-Nucleotidase from the venom Hemachatus haemachates	E.C. 3.1.3.5	11 enzyme units, at pH 5.5	108	1	3.04×10^{-5}	Homogeneous
			15 enzyme units, at pH 8.5	120	2[b]	—	Homogeneous

[a] Representative experiments carried out at the Institute of Biochemistry, Uppsala, Sweden. Dashes are used where data were not given.
[b] Possibly due to enzyme denaturated during electrophoresis.

Fig. 3. Free zone electrophoretic patterns. (a) Amniotic fluid from a normal pregnancy in the second trimester (21st week); (b) maternal plasma. A, B, C, and D are the main electrophoretic fractions in amniotic fluid. Arrows indicate the positions of protein zones at the start. +, anode; −, cathode. [From Jonasson and Hjertén (1973). Reproduced by permission of *Acta Obstet. Gynec. Scand.* Almqvist & Wiksell Periodical Company.]

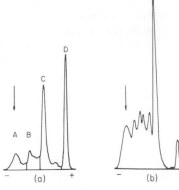

on Protein Mixtures and Pure Enzyme Preparations

			Compared with other methods			
Molecular weight	Isoelectric point	Specific activity	Purification factor	$A^{1\,cm}_{280\,nm}$	Molecular extinction coefficient	Ref.
						Hjertén (1967a,b, 1970)
						Hjertén (1970), Jonasson and Hjertén (1973)
						Jonasson and Hjertén (1973)
						Hjertén (1970)
						Hjertén (1967a, b)
22,000	—	1700 U/mg	250–425 x	18.6	44.4×10^3	Hofsten *et al.* (1965), Hjertén (1967a,b)
						Hjertén (1970)
6,787	—	—	—	—	8.7×10^3	Karlson *et al.* (1966)
60,000	4.0	Relative	115 x	30.0	—	Janson and Porath (1966), Hjertén (1970)
18.1 S (1,000,000)	—	Relative	49 x	—	—	Brishammer and Philipson (1966)
47,100	6.10	—	—	—	—	Squire *et al.* (1964)
47,000	5.74 (10°C)	7200 U/mg	60 x	—	—	Berghem and Pettersson (1974)
—	—	Relative	2000 x	—	—	Björk (1964, 1967)

the purity of isolated proteins, especially (microbial) enzymes. Table II lists representative experiments.

Although the equipment for free zone electrophoresis is commercially manufactured by Incentive Research & Development, A. B., Bromma, Sweden, all experiments summarized in Table II were conducted at the Institute of Biochemistry at the University of Uppsala.

The free zone electrophoresis method is very well established theoretically as well as experimentally, according to Hjertén (1967a,b). However, the separational ability of this method appears average when analyzing complex mixtures of proteins (Table II).

Comparing two entirely different methods, it may appear critical to state that analytical isotachophoresis has a much higher resolving power for complex mixtures than free zone electrophoresis. The Tachophor apparatus* thus shows a total of 37 major and minor components in normal human serum, and at least 16 components in the γ globulin fraction of human cerebrospinal fluid (Kjellin et al., 1975).

V. FREE ZONE ISOELECTRIC FOCUSING[†]

Free zone isoelectric focusing is performed with the free zone electrophoresis equipment developed by Hjertén (1967a,b), with slight modifications. Approximately 0.1 mg of protein can be isofocused in the slowly rotating quartz tube in 30 hr. There is a serious risk of artifactual ultraviolet absorbing zones ("false peaks") which arise from possible impurities in the Ampholine carrier ampholytes solutions; these impurities tend to form white precipitates, which are normally observed in blank experiments. Figure 4 illustrates the isoelectric focusing of human serum albumin. The isoelectric points measured were peak II, 5.38 (5.19) at 23°C (4°C); peak III, 6.50 (6.30) at 23°C (4°C).

The resolving power does not appear very high for separated zones. Free zone isofocusing of carboxyhemoglobin (Lundahl and Hjertén, 1973) resulted in a difference of main focused zones of 0.08 pH unit. However, for polyacrylamide gel isoelectric focusing, a difference of 0.02 pH unit between focused bands was observed (Bours, 1971, 1973a). It appears that the incorporation of the isoelectric focusing principle into the free zone electrophoresis system essentially does not increase the degree of separation (Fig. 4). Because of the cost of the apparatus, the required technical skill, and the apparently lower separational power of the free zone isoelectric focusing when compared to thin-layer isoelectric focusing, employing this system in my opinion is not justified. Currently, only the former method has been applied by Lundahl and Hjertén (1973).

---

* LKB 2127, Bromma, Sweden.
[†] Lundahl and Hjertén (1973).

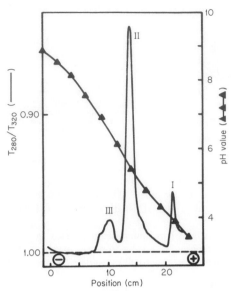

Fig. 4.   Free zone isoelectric focusing. Sample contents about 0.10 mg of human serum albumin. The tube is scanned in ultraviolet light, and the ratio of the transmissions at wavelengths 280 and 320 nm is recorded as the ratio T_{280}/T_{320}. Peaks II and III represent proteins. The pH of the material corresponding to peak II was 5.38 $\pm$ 0.05 at 23°C. Peak I is a "false" peak of a white precipitate coming from the Ampholine used. [From Lundahl and Hjertén (1973). Reproduced by permission of the *Annals of the New York Academy of Sciences*.]

VI.   FREE ISOELECTRIC FOCUSING*

A.   Description of the Method

Using a modification of the apparatus described by Macko and Stegemann (1970), isoelectric focusing of a previously purified protein is performed in a 1- or 2.5-m polythylene tubing. Extensive and accurate descriptions have been published (Bours, 1973c). The tubing (4 mm i.d., 6 mm o.d., 2.5 m length) is marked every 2.5 cm, and filled with 48–50 ml of a solution containing 2% Ampholine carrier ampholytes of the desired pH range. To eliminate all possible air bubbles, both sides are stoppered, and the tubing is fixed in close turns around a Perspex axis (Bours, 1973c) or around a copper pipe, as shown in Fig. 5. The protein samples, ranging in weight from 5 to 50 mg, are dissolved in 1–2 ml distilled water and poured in. The samples are inserted at the anodic site (Valmet, 1970a; Bours, 1973a; Hemmings and Jones, 1974), where precipitation of albumin, α-crystallin, or any other protein with lower isoelectric point—such as β-lactoglobulin (Radola, 1973)—may be expected. The electrode liquids are 0.5 ml 0.1% phosphoric acid filled at the anodic site, and 0.5 ml 0.4% ethanolamine at the cathodic site. Platinum electrodes are

* According to Macko and Stegemann (1970).

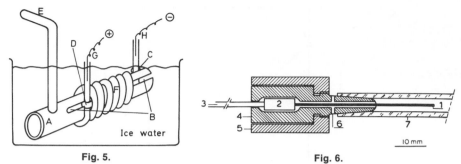

Fig. 5.   **Fig. 6.**

Fig. 5.   Apparatus for free isoelectric focusing of proteins. A, brass tube, 25 × 170 mm B, slit, 4 × 120 mm; C, fixed slot for tube F; D, movable slot for tube F; E, supporting bar; F, polyethylene tube (4 mm i.d., 6 mm o.d., length 1 m); G and H, platinum electrodes. [From Chilla *et al.* (1973).]

Fig. 6.   Schematic drawing of platinum electrode used in free isoelectric focusing. 1, platinum wire; 2, soldered connection; 3, cable to power supply; 4, Perspex holder; 5, Perspex protective collar (dimensions 18 × 25 mm); 6, gas outlet; 7, polyethylene tubing. [From Bours (1973c).]

specially designed for safety measures, according to the schematic drawing shown in Fig. 6. These electrodes are pushed inside the free ends of the coiled tubing and an electric field of 3800 V and 850–160 μA is applied for 27 hr. Figure 7 shows the time course of current during free isoelectric focusing.

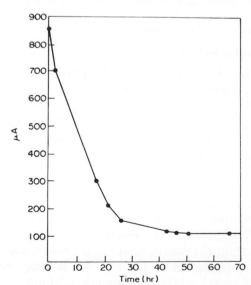

Fig. 7.   Time course of current in microamperes during free isoelectric focusing. Electric field of 3800 V and 850–160 μA for 27 hr, until the microamperage reaches a constant value of about 100 μA. Initial power 6.5 W, final 0.4 W. [From Bours (1973c).]

After a minimum of 27–30 hr, the electrodes are removed and the coil is stoppered and placed in the refrigerator at $-20°C$ until frozen. The contents of the frozen fractions are placed into 100 small test tubes. All samples, containing approximately 0.5 ml, can be assayed for pH, absorption at 280 nm, immunoelectrophoresis (Fig. 10), enzyme activity, or submitted to isoelectric refocusing on a thin-layer isoelectric focusing gel plate (Fig. 11).

B.   Comparison of Various Results

A number of successful purifications have been carried out by various authors, and are summarized in Table III. Figures 8 and 9 show the determination of pH versus enzyme activity for glucose-6-phosphate dehydrogenase (Chilla et al., 1973) and ribose phosphate isomerase (Domagk et al., 1973), respectively, which were highly purified by free isoelectric focusing. Further evidence for homogeneity of purified enzyme preparations is shown in Table III (Numbers 3, 4, 5) under the headings of specific activity, purification factor, molecular weight, and isoelectric point.

C.   The Isolation of δ-Crystallin from Chick Iris*

A number of antigenic constituents of the eye lens are also present in other intraocular tissues (Bours, 1973a, 1974). The cross-reactivity of the antigens of the lens with similar antigens of other intraocular tissues such as the iris has also been shown by Bours (1973a,b, 1974). One of the most important proteins in the chick lens is δ-crystallin. This crystallin is of embryonic origin and serves as the main constituent of the chick embryonic lens (Rabaey, 1962; Truman et al., 1972). In later stages of development, the concentration of δ-crystallin appears to be about 20% (Bours, 1974), while the percentage in the iris is only about 3%.

The rationale of this free isofocusing experiment was to isolate low concentrations of δ-crystallin from the abundance of other crystallins and serum proteins present in chick iris (Bours, 1973a,b). The material used for this isolation was 30 mg of chick iris extract, prepared as described earlier (Bours, 1973a). Thus, a δ-crystallin is isolated from iris which is immunologically identical to the δ-crystallin from lens, while identical isoelectric points of the various components from lens and iris are measured. Following free isofocusing, the pH gradient is measured. As determined by immunoelectrophoresis of chick lens crystallins using polyvalent antiserum, the fractions 20–24 contained δ-crystallin in a pure form (Figs. 10b–f). The pH of these liquid fractions measured 5.30–5.45. For unknown reasons, these pH values are slightly higher than anticipated (Table IV) from the isoelectric points of

* Bours (1973c).

TABLE III   Free Isoelectric Focusing According to Macko and

No.	Protein(s)	Quantity analyzed	Ampholine concentration (%)	Sucrose added (%)	Temperature during run (°C)	Voltage (V)	Amperage (μA)	Period	No. of fractions	Length of tubing (cm)	$\partial pH/\partial x$ for 1 pH unit/cm length of tubing
1	Potato tuber proteins	10–50 mg	1	10	2	600; 1000	160; 50	12 hr;	40	100	0.037
2	Eye lens γ-crystallins	30 mg	2	10	4	3800	850	27 hr	60	150	0.038 0.038
3	Glucose-6-phosphate dehydrogenase from *Candida utilis* (E.C. 1.1.1.49)	2 ml	2	None	0	1000	200	3 days	40	100	0.030 0.030
4	Ribose phosphate isomerase from *Candida utilis* (E.C. 5.3.1.6)	29 mg	2	None	0	1000	400	67 hr	50	100	0.033
5	Ribose phosphate isomerase from skeletal muscle (E.C. 5.3.1.6)	—	2	None	0	1000	400	46 hr	50	100	—
6	Phosphoryl-phosphatase from rabbit serum	10–70 mg	2	None	0	1000	50	90 hr	100	100	0.044 0.180
7	Diisopropyl-phosphorofluoridate fluorohydrolase from *Escherichia coli* (E.C. 3.8.2.1.)	—	2	None	0	1000	50	90 hr	100	100	0.050 0.140

[a] Representative experiments carried out at the Physiological Chemistry Institute of the University of Göttingen, West Germany. Dashes are used where data were not given.

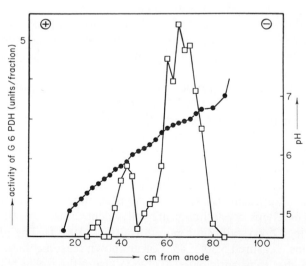

Fig. 8.   Free isoelectric focusing of glucose-6-phosphate dehydrogenase. Aliquots of each of the 40 fractions obtained were assayed for glucose-6-phosphate dehydrogenase activity (open squares) and pH (black dots). Isoelectric points are 5.50, 5.87, and 6.54. [From Chilla *et al.* (1973).]

tegemann (1970) on Protein Mixtures and Pure Enzyme Preparations

pH region where $\partial pH/\partial x$ measured	Purity	Isoelectric point(s)	Molecular weight	Specific activity (U/mg)	Purification factor (mg)	Recovery (%)	Ref.
6–9	Homogeneous	6.2–8.7	10,200 13,800	—	—	—	Macko and Stegemann (1970), Stegemann et al. (1973)
6–7 7–8	Two homogeneous fractions	6.10–6.90 7.25; 7.35; 7.60; 7.80	—	—	12 x	27 (46)	Bours (1973c)
5–6 6–7	Homogeneous	5.50; 5.87; 6.54	110,000	510	2850 x	46	Chilla et al. (1973)
4–5	Homogeneous	4.7	105,000	356	224 x	27	Domagk et al. (1973)
—	Homogeneous	5.9	228,000	355	3070 x	37	Domagk et al. (1974)
5–6 6–7	Partially purified	5.7; 6.3	500,000	—	70 x	—	Zech and Zürcher (1973, 1974)
5–6 6–7	Partially purified	5.3; 5.7; 6.1; 7.8	—	—	—	—	Zech and Wigand (1975)

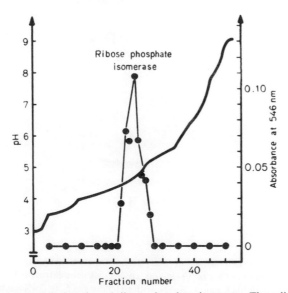

Fig. 9.   Free isoelectric focusing of ribose phosphate isomerase. The solid line represents the pH gradient. The ribose phosphate activity (•–•) is given by the absorbance at 546 nm in a 1-cm cell. The isoelectric point of the enzyme is 4.7. [From Domagk et al. (1973). Reproduced by permission of Springer-Verlag, Inc.]

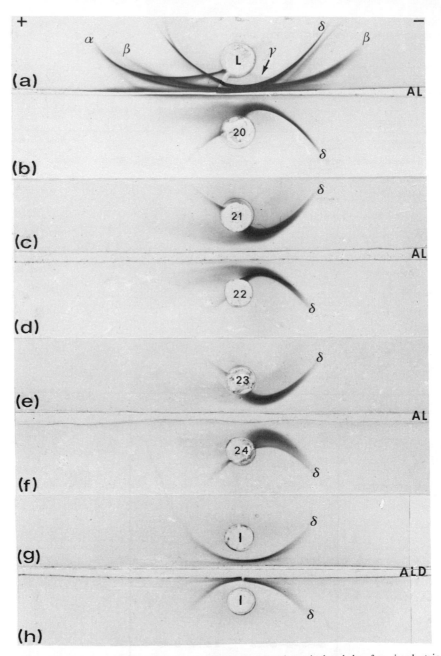

Fig. 10. Immunoelectrophoresis of δ-crystallin fractions isolated by free isoelectric focusing of chick iris extract (Bours, 1973c). (a) Total chick lens extract (L). (b)–(f) Chick iris δ-crystallin detected in liquid fractions 20–24, with pH values of 5.30–5.45. Each fraction shows one characteristic precipitin line developed with a polyvalent antiserum to chick lens crystallins (AL). (g), (h) Chick iris extract (I), showing one precipitin line when developed with a monospecific antiserum (ALD) directed to chick lens δ-crystallin (Bours, 1973a). α, α-crystallin; β, β-crystallin; γ, γ-crystallin (Bours, 1973e); δ, δ-crystallin.

TABLE IV

Isoelectric Points and Separational Width of δ-Crystallin Components
Isolated from Chick Iris Extract[a]

| δ-Crystallin component No. | Chick lens extract | Chick iris extract | Free isoelectric focusing fraction, No. | | | | | $\partial pH/\partial x$ |
			20	21	22	23	24	
1	5.18	5.18	5.18	5.18	—[b]	—	—	0.010
2	5.20	—	5.20	5.20	5.20	—	—	0.013
3	5.23	5.23	5.23	5.23	5.23	—	—	0.013
4	5.26	—	5.26	5.26	5.26	5.26	5.26	0.012
5	5.28	5.28	—	—	5.28	5.28	5.28	0.006
6	5.30	—	—	—	5.30	5.30	5.30	0.006
7	5.34	5.34	—	—	5.34	5.34	5.34	0.006
			(5.30)[c]	(5.35)	(5.40)	(5.42)	(5.45)	

[a] Measured at 4°C.
[b] Absent.
[c] Numbers in parentheses are the pH values of the free isofocusing fractions; Fig 10.

the isolated proteins (cf. Bours, 1973c). These fractions were submitted to thin-layer isoelectric focusing by Bours (1971) (Fig. 11, 20–24), and were compared with chick lens crystallins (Fig. 11a) as well as proteins from chick iris (Fig. 11b). It appears that the δ-crystallin from chick iris has an incomplete number of components (Bours, 1974) (Fig. 11b), but nevertheless contained in this enriched form were all seven components which were originally present in iris, and which are, as a matter of fact, also present in the lens (Bours, 1971). The purity of this preparation was assayed by both immunoelectrophoresis and thin-layer isoelectric focusing.

Table IV gives isoelectric points determined at 4°C. This is in agreement with the work of Piatigorsky *et al.* (1973, 1974) who estimated the isoelectric point of chick lens δ-crystallin (5.1) by density gradient isoelectric focusing.

According to these experiments, which accomplished the quantitative isolation of δ-crystallin from chick iris, it should be concluded that the crystallins are eye-specific and not lens-specific, as assumed (Uhlenhuth, 1903).

D. The Value of the Method

Employing coils of polyethylene tubing, free isoelectric focusing is a very efficient method for the purification of proteins on a semipreparative scale. The application of this method requires only a high-voltage power source for direct current, which is probably available in most laboratories. The degree

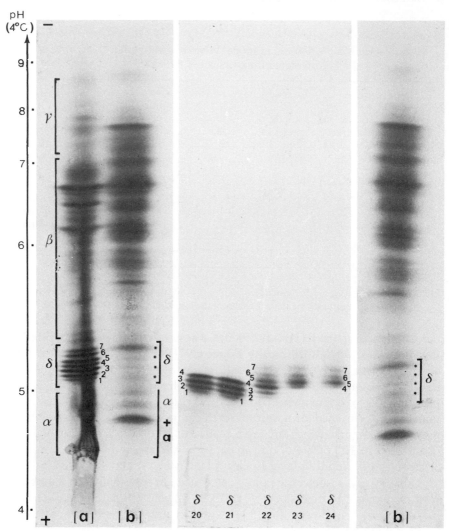

Fig. 11. Thin-layer isoelectric focusing of δ-crystallins according to Bours (1971). Ten-microgram samples of δ-crystallin from fractions 20–24 isolated by free isoelectric focusing of chick iris extract. (a) 0.2 mg total chick lens extract; (b) 0.6 mg chick iris extract. Numbers 1–7 are the δ-crystallin components detected in chick lens (a) and after free isofocusing of chick iris in fractions 20–24. The asterisk denotes each of the four components present in chick iris extract. α, α-crystallin; β, β-crystallin; γ, γ-crystallin (Bours, 1973e); δ, δ-crystallin; a, albumin. The scale shows the pH values along the gel.

of focusing and the resolving power of this method can be favorably influenced according to Vesterberg (1968) by

1.   increasing the field strength, i.e., the potential applied to the coil, thereby obtaining better resolution;
2.   increasing the column length, thus obtaining better resolution, and selecting Ampholine carrier ampholytes having pI values in a limited pH range, thereby obtaining a low value of $\partial pH/\partial x$.

In the free isofocusing experiments reported (Table III) the value of $\partial pH/\partial x$ has an average of 0.04 pH unit/cm at the level of the protein zones. For proteins isolated by free isoelectric focusing the mean of this value is calculated to be 0.01 for chick δ-crystallin (Table IV), and 0.02 for bovine γ-crystallin (Bours, 1973c). In density gradient isoelectric focusing, the excellent resolution of well-separated zones is lost during elution of the column. This drawback can be eliminated in the free isofocusing method because the liquid contents of the coil are frozen directly after the experiment, thus avoiding diffusion which broadens the separated protein zones.

The free isoelectric focusing method is preeminently suitable for purifying enzymes, giving preparations with high specific activities as can be concluded from the data shown in Table III, numbers 3, 4, and 5.

VII.   THE MICROHETEROGENEITY OF PROTEINS

A.   Synthetic Heterogeneity of δ-Crystallin

The microheterogeneity of the chick δ-crystallin (Bours, 1974) is probably an example of synthetic heterogeneity (Awdeh et al., 1970; Williamson et al., 1973). This microheterogeneity is attributed to slight differences in the primary structure of each of the seven crystallin components (Bours, 1974). The isoelectric points of the components 1–7 (Fig. 11, Table IV) are very close to one another. The difference in the isoelectric point of each component is 0.02–0.05 of a pH unit, and the group of seven components ranges in isoelectric point, by only 0.16 of a pH unit (Table IV) (Bours, 1974).

Although it is impossible to calculate the isoelectric points of the δ-crystallin components because the number of titratable groups and the amino acid composition are not known, it is still possible to account for the shifts in pI from one component to the next. For this purpose, Stig Fredriksson has calculated theoretically these differences in isoelectric points between the seven components of δ-crystallin, according to the Linderstrøm–Lang equation mentioned by him (Fredriksson, 1975). From the presumed stepwise deamination of 6 × one asp·NH$_2$ or glu·NH$_2$ residue into β,γ-carboxyl groups from component 7 through 1, or from the removal of 6 × one basic group, these shifts in pI are calculated. When 85 β,γ-carboxyl groups in component 7 are assumed, then 91 of these groups should be present in component 1. Further, when 5.34, the isoelectric point of the seventh component, is taken as reference, the values of the isoelectric points for the seven components are 5.34, 5.31, 5.28, 5.25, 5.22, 5.20, and 5.18. These theoretic values are in close agreement with the experimental data in Table IV.

The δ-crystallin components share:

1. the same immunological determinant(s) (Bours, 1974);
2. the same molecular weight of 155,000–165,000 (Zwaan, 1968).

These facts indicate that the δ-crystallin components are very similar in polypeptide chain composition, and are encoded by gene duplication (Williamson *et al.*, 1973).

B. Postsynthetic Heterogeneity of Monoclonal Immunoglobulin G

The microheterogeneity of human monoclonal IgG (Mulder and Verhaar, 1973; Brendel *et al.*, 1974) is an example of postsynthetic heterogeneity (Awdeh *et al.*, 1970; Williamson *et al.*, 1973), and is reproduced in Fig. 12.

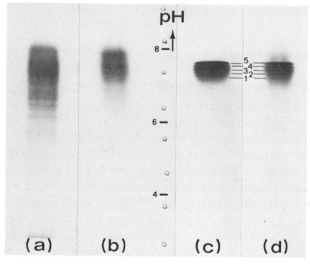

Fig. 12. Thin-layer isoelectric focusing of human immunoglobulin G according to Bours (1971). (a) 0.2 mg normal human IgG (Nordick, Tilburg, The Netherlands); (b) 0.2 mg polyclonal IgG; (c) 0.2 mg monoclonal IgG. (b) and (c) are prepared by fractionation with DEAE-Sephadex A-50 column chromatography; (d) 0.2 mg of a mixture of polyclonal and monoclonal IgG, prepared by batch absorption with DEAE-Sephadex A-50. The scale shows the pH values along the gel. Numbers 1–5 represent isofocused components with isoelectric points of 7.50, 7.60, 7.69, 7.79, and 7.86, respectively (Mulder and Verhaar, 1973). [From Brendel *et al.* (1974). Reproduced by permission of Elsevier Publishing Co.]

The isoelectric points of the five components are 7.50, 7.60, 7.69, 7.79, and 7.86. These components differ by 0.07–0.10 of a pH unit and range in isoelectric points by 0.36 of a pH unit. The discrete banding pattern of five isofocused IgG components from myeloma proteins results from a postsynthetic charge alteration by deamidation of a biosynthetically homogeneous protein (Brendel *et al.*, 1974; Williamson *et al.*, 1973).

NOTE

Recently, an informative survey article was published by S. Hjerten (1976) on free zone electrophoresis. A comparison is made therein between carrier-free electrophoresis, isoelectric focusing, and isotachophoresis, as carried out with the free zone electrophoresis equipment.

ACKNOWLEDGMENTS

I thank Dr. A. A. Swanson, Associate Professor of Biochemistry and Ophthalmology, for useful discussions, and I thank Prof. Dr. rer. nat. O. Hockwin for reviewing the manuscript. Thanks are also extended to Mr. Th. Hulskes (University of Utrecht) and to Miss Barbara Polenz for taking photographs and for reproducing the illustrations.

REFERENCES

Awdeh, Z. L., Williamson, A. R., and Askonas, B. A. (1970). *Biochem. J.* **116**, 241–248.
Berghem, L. E. R., and Pettersson, L. G. (1974). *Eur. J. Biochem.* **46**, 295–305.
Björk, W. (1964). *Biochim. Biophys. Acta* **89**, 483–494.
Björk, W. (1967). *Ark. Kemi* **27**, 555–569.
Bours, J. (1971). *J. Chromatogr.* **60**, 225–233.
Bours, J. (1972). *Ophtal. Res.* **3**, 9.
Bours, J. (1973a). *Exp. Eye Res.* **15**, 299–319.
Bours, J. (1973b). *Exp. Eye Res.* **16**, 487–499.
Bours, J. (1973c). *Exp. Eye Res.* **16**, 501–515.
Bours, J. (1973d). *Sci. Tools* **20**, 29–34.
Bours, J. (1973e). *Exp. Eye Res.* **17**, 403.
Bours, J. (1974). *Doc. Ophthalmol.* **37**, 1–46.
Bours, J. (1975). *In* "Progress in Isoelectric Focusing and Isotachophoresis" (P. G. Righetti, ed.), pp. 235–256. Elsevier, Amsterdam.
Brendel, S., Mulder, J., and Verhaar, M. A. T. (1974). *Clin. Chim. Acta* **54**, 243–248.
Brishammer, S., and Philipson, L. (1966). *Biochim. Biophys. Acta* **127**, 140–150.
Catsimpoolas, N. (1973). *Separ. Sci.* **8**, 71–121.
Chilla, R., Doering, K. M., Domagk, G. F., and Rippa, M. (1973). *Arch. Biochem. Biophys.* **159**, 235–239.
Denckla, W. D. (1974). *J. Clin. Invest.* **53**, 572–581.
Domagk, G. F., Doering, K. M., and Chilla, R. (1973). *Eur. J. Biochem.* **38**, 259–264.
Domagk, G. F., Alexander, W. R., and Doering, K. M. (1974). *Hoppe-Seyler's Z. Physiol. Chem.* **355**, 781–786.
Fawcett, J. S. (1975). *In* "Isoelectric Focusing" (J. P. Arbuthnott and J. A. Beeley, eds.), pp. 23–43. Butterworths, London.
Fredriksson, S. (1975). *J. Chromatogr.* **108**, 153–167.
Haglund, H. (1967). *Sci. Tools* **14**, 17–23.

Haglund, H. (1971). *Methods Biochem. Anal.* **19**, 1–104.

Hemmings, W. A., and Jones, R. E. (1974). *Immunology* **27**, 343–350.

Hjertén, S. (1967a). Thesis, Univ. of Uppsala, Almqvist and Wiksells, Uppsala, Sweden. pp. 1–117. Uppsala.

Hjertén S. (1967b). *Chromatogr. Rev.* **9**, 122–219.

Hjertén S. (1970). *Methods Biochem. Anal.* **18**, 55–79.

Hjerten, S., (1976). *In* "Methods of Protein Separation" (N. Catsimpoolas, ed.) Vol. 2, pp. 219–232. Plenum, New York.

Hofsten, B. V., van Kley, H., and Eaker, D. (1965). *Biochim. Biophys. Acta* **110**, 585–598.

Janson, J. -C., and Porath, J. (1966). *Methods Enzymol.* **8**, 615–621.

Jonasson, L. -E., and Hjertén, S. (1973). *Acta Obstet. Gynecol. Scand.* **52**, 345–354.

Karlsson, E., Eaker, D. L., and Porath, J. (1966). *Biochim. Biophys. Acta* **127**, 505–520.

Kiryukhin, I. F. (1972). *Biull. Eksp. Biol. Med.* **74**, 120–122.

Kjellin, K. G., Moberg, U., and Hallander, L. (1975). *Sci. Tools* **22**, 3–7.

Lundahl, P., and Hjertén, S. (1973). *Ann. N.Y. Acad. Sci.* **209**, 94–111.

Macko, V., and Stegemann, H. (1970). *Anal. Biochem.* **37**, 186–190.

Mulder, J., and Verhaar, M. A. T. (1973). *Int. Res. Commun. Syst.* (73-11), 17-22-4.

Piatigorsky, J., Rothschild, S. S., and Milstone, L. M. (1973). *Develop. Biol.* **34**, 334–345.

Piatigorsky, J., Zelenka, P., and Simpson, R. T. (1974). *Exp. Eye Res.* **18**, 435–446.

Rabaey, M. (1962). *Exp. Eye Res.* **1**, 310–316.

Radola, B. J. (1973). *Ann. N.Y. Acad. Sci.* **209**, 127–143.

Squire, P. G., Delin, S., and Porath, J. (1964). *Biochim. Biophys. Acta* **89**, 409–421.

Stegemann, H., Francksen, H., and Macko, V. (1973). *Z. Naturforsch.* **28**, 722–732.

Svensson, H. (1948). *Advan. Protein Chem.* **4**, 251–268.

Talbot, P. (1975). *In* "Isoelectric Focusing" (J. P. Arbuthnott and J. A. Beeley, eds.), pp. 270–274. Butterworths, London.

Talbot, P., and Caie, I. S. (1975). *In* "Isoelectric Focusing" (J. P. Arbuthnott and J. A. Beeley, eds.), pp. 74–77. Butterworths, London.

Tolkacho, N. V. (1974). *Ukr. Biokhim.* **46**, 441–445.

Troitski, G. V., Kiryukhin, I. F., Tolkacheva, N. V., and Azhitsky, G. Y. (1974a). *Vop. Med. Khim.* **20**, 24–31.

Troitskii, G. V., Zaryalov, V. P., and Abramov, U. M. (1974). *Dokl. Akad. Nauk USSR* **214**, 955–958.

Truman, D. E. S., Brown, A. G., and Campbell, J. C. (1972). *Exp. Eye Res.* **13**, 58–69.

Uhlenhuth, P. T. (1903). *In* "Festschrift zum 60-sten Geburtstag von Robert Koch," pp. 49–74. Fischer, Jena.

Valmet, E. (1968). *Sci. Tools* **15**, 8–11.

Valmet, E. (1969). *Sci. Tools* **16**, 8–13.

Valmet, E. (1970a). *Protides Biol. Fluids Proc. Colloq.* **17**, 401–407.

Valmet, E. (1970b). *Protides Biol. Fluids Proc. Colloq.* **17**, 443–448.

Vesterberg, O. (1968). Thesis, Univ., of Stockholm, Stockholm, Sweden.

Vesterberg, O., and Svensson, H. (1966). *Acta Chem. Scand.* **20**, 820–834.

Williamson, A. R., Salaman, M. R., and Kreth, H. W. (1973). *Ann. N.Y. Acad. Sci.* **209**, 210–224.

Zech, R., and Wigand, K. D. (1975). *Experientia* **31**, 157–158.

Zech, R., and Zürcher, K. (1973). *Life Sci.* **13**, 383–389.

Zech, R., and Zürcher, K. (1974). *Arzneim. Forsch.* (*Drug Res.*) **24**, 337–340.

Zwaan, J. (1968). *Exp. Eye Res.* **7**, 461–472.

9 TRANSIENT STATE ISOELECTRIC FOCUSING

Nicholas Catsimpoolas

Biophysics Laboratory
Department of Nutrition and Food Science
Massachusetts Institute of Technology
Cambridge, Massachusetts

I. INTRODUCTION

A. Principle

Transient state isoelectric focusing (TRANS-IF) involves the study of the kinetic behavior of charged amphoteric molecules during their transport by the electric field in an isoelectric focusing system (Catsimpoolas, 1973a,b). This is achieved by repetitive optical scanning of the concentration distribution of migrating solutes followed by computer analysis of the peaks

(Catsimpoolas, 1975a). The scanning of the separation path is usually carried out by absorbance measurements although other optical methods such as fluorescence and refraction can be used. Upon completion of the experiment, the various scans obtained as a function of time can be utilized in producing kinetic information pertinent to methodological aspects of the method and to the estimation of physical constants.

The TRANS-IF method was developed by the author in 1973 (Catsimpoolas, 1973a,b) based on earlier experiments (Catsimpoolas and Wang, 1971; Catsimpoolas, 1971a) using scanning isoelectric focusing. Variations of the latter technique were also reported two years later by Rilbe and co-workers (see Rilbe, 1975) and Lundahl and Hjertén (1973). Both poly-acrylamide gels and density gradients can be utilized as the supporting medium for the separation.

B. Transient and Steady States

In an attempt to define some of the processes that occur during an iso-electric focusing experiment, a simple classification of transient and steady states is depicted in Table I. The carrier ampholyte species are considered separately from the proteins. This separation is favored by the fact that the formation of the pH gradient is a cooperative phenomenon requiring the presence of a large number of ampholyte species, whereas proteins can be focused as single species in an ampholyte system. In addition, carrier

TABLE I

Transient Steady States

Stage	State	Remarks
Focusing	TRANS-1	Focusing of ampholytes
	TRANS-2	Focusing of proteins
Steady state	SS-1	Steady state of ampholytes
	SS-2	Steady state of proteins
Deformation	TRANS-3	Deformation of the SS-1 state
	TRANS-4	Deformation of the SS-2 state
Defocusing	TRANS-5	Diffusional defocusing of ampholytes[a]
	TRANS-6	Diffusional defocusing of proteins[a]
Refocusing	TRANS-7	Refocusing of ampholytes[b]
	TRANS-8	Refocusing of proteins[b]
Refocusing steady state	SS-3	Steady state of ampholytes
	SS-4	Steady state of proteins

[a] In the absence of the electric field.
[b] After reapplication of the electric field.

ampholytes and proteins exhibit widely different molecular size and conductance. Thus, the kinetics of focusing of ampholytes, especially in polyacrylamide gels, cannot be considered as similar to the focusing of proteins. The terminology given in Table I is now discussed in more detail.

The TRANS-1 state represents the formation of the pH gradient by the carrier ampholytes. This is accomplished by the electrophoretic migration of the individual carrier ampholyte species (A_1, A_2, ..., A_n) toward their respective isoelectric points (pI_1, pI_2, ..., pI_n). The kinetics of focusing of the ampholytes are very complex and depend on a number of factors. First, the nature of the ampholytes themselves has to be considered, i.e., their initial bulk concentration, number of species, pI distribution, relative concentration of individual species, the pH–mobility curve of each species, and their size. In addition, the "mode of loading" of the ampholytes, which can be uniform, gradient positive, or gradient negative (Table II), will influence the kinetics of focusing. When a density gradient is used there is a "gradient viscosity" effect

TABLE II

Terminology Relating Carrier Ampholyte Mode of Loading, Sample Position, and Polarity Orientation–Density Gradient Relationship

Terminology	Abbreviation	Explanation
Density gradient		
1. Positive	DGP	Density increases toward positive electrode
2. Negative	DGN	Density increases toward negative electrode
Carrier ampholyte loading		
1. Uniform	AUN	Uniform concentration distribution
2. Gradient (positive)	AGP	Concentration increases toward positive electrode
3. Gradient (negative)	AGN	Concentration increases toward negative electrode
Sample position		
1. Uniform	UN	Uniform concentration distribution
2. Gradient (positive)	GP	Concentration increases toward positive electrode
3. Gradient (negative)	GN	Concentration increases toward negative electrode
4. Pulse (positive)	PP	Zone on the positive end of the column
5. Pulse (negative)	PN	Zone on the negative end of the column
6. Pulse (both ends)	PB	Zone on both the positive and negative ends
7. Pulse (pI positive)	PIP	Zone between expected pI position and the positive electrode
8. Pulse (pI negative)	PIN	Zone between expected pI position and the negative electrode
9. Pulse (pI)	PPI	Zone at the expected pI position

which depends on the concentration of sucrose in the gradient and the temperature. In addition, the position of the electrode polarity (positive or negative) in respect to the dense part of the gradient will affect the electrophoretic migration of the individual ampholyte species. The effect of polyacrylamide gels is expected to be mainly due to the differential sieving of the carrier ampholytes. Gels of variable %T, %C values retard ampholyte migration to a different extent; thus, there may be a delay in the formation of the pH gradient in restrictive pore gels. Even at fixed %T, %C gels, any polydispersity of ampholytes in respect to molecular weight distribution may introduce an additional variable. From the aforementioned, it should be obvious that there is no easy answer to the question most often asked; i.e., how long does it take for the pH gradient to be formed. Should this be found empirically, it would only apply to a fixed set of experimental conditions and batch of ampholytes.

Evaluation of the kinetics of the TRANS-1 state by TRANS-IF can be carried out by using ultraviolet absorbing amphoteric compounds such as hystidyltyrosine, tyrosyl glutamic acid, and lysyltyrosine (Catsimpoolas and Campbell, 1972; Catsimpoolas *et al.*, 1974a,b). A mixture of such compounds covering approximately equidistant points in the pH gradient should provide valuable information concerning the kinetics of focusing.

Focusing of proteins (TRANS-2 state) involves the electrophoretic migration of the individual protein species toward their respective p*I* values. This process can occur either in a field of prefocused carrier ampholytes (SS-1/ TRANS-2) or by concurrent focusing of carrier ampholytes and proteins (TRANS-1/TRANS-2). In the simplest SS-1/TRANS-2 case the combination of "density gradient mode" and "sample position" can produce several kinetic variations of the TRANS-2 state. More complex variations can be expected in the TRANS-1/TRANS-2 system for a given protein. Again, this emphasizes the necessity of defining of the exact conditions of any isoelectric focusing experiment where kinetic data (e.g., minimal focusing time) are to be compared or interpreted. In addition, to these considerations, other factors (e.g., gel sieving, concentration of carrier ampholytes, and species distribution) will undoubtedly contribute to the kinetic aspects of the TRANS-2 state. Furthermore, we should realize that even within a well-defined isoelectric focusing system there are as many TRANS-2 states as there are proteins in the sample because each protein exhibits its own characteristic pH–mobility curve.

The SS-1 state is achieved when all the carrier ampholyte species are at the "steady state," i.e., focused. Since it is doubtful that all the ampholyte species reach the steady state simultaneously, some of the ampholyte species can be at the TRANS-1 state while others have reached their individual steady state.

Again in the case of deformation (see below TRANS-5 state), the TRANS-1 and TRANS-5 states can occur simultaneously without ever really achieving the ideal SS-1 state. It should be mentioned that the SS-1 state as described here does not refer to the stable pH gradient but rather to the steady state of the ampholyte species as described by the steady-state equation (Svensson, 1961) which takes into account only the statistical moments of the peak (e.g., constant variance).

The SS-2 state denotes the steady state of a protein species, i.e., focused. The SS-2 state can exist only in the presence of the SS-1 state. The requirement is that all the statistical moments of the protein distribution remain constant as long as the electric field is applied. Again, pH is not implicated; in other words, a focused protein can maintain constancy of pI without being at the SS-2 state. In addition, there are as many SS-2 states as there are proteins in the sample. This means that within an experiment certain proteins may have reached the SS-2 state while others are still in the TRANS-2 state. In particular cases, e.g., in the presence of sieving, certain proteins may never attain the SS-2 state even if an SS-1 state can be sustained. This latter condition can be a serious source of artifacts.

When the SS-1 state is not stable, it drifts into the TRANS-3 state. This phenomenon can occur for a number of reasons, including electroosmosis, thermal convection, leaking of ampholytes into the electrodes, and decomposition products at the electrodes. The net result is that the SS-1 state is abolished. In some instances a direct transition from the TRANS-1 into the TRANS-3 state can occur. Deformation of the protein steady state (TRANS-4) can be the result of either the abolishment of the SS-1 state (TRANS-3) or changes in the protein (e.g., association, dissociation, aggregation) occurring at the isoelectric point. Again under certain conditions a protein may pass from the TRANS-2 state directly into the TRANS-4 state without ever achieving the SS-2 state.

The TRANS-5 state involves the diffusion of ampholytes in the absence of the electric field after the SS-1 (or TRANS-3) state has been achieved. The interruption of the electric field is performed intentionally in order to follow the diffusion of focused compounds and, thus, be able to measure their apparent diffusion coefficients. The TRANS-7 state occurs when the electric field is reapplied after the TRANS-5 state. Each ampholyte species is expected to be refocused from a normal concentration distribution (Catsimpoolas *et al.*, 1974b). Note that this is different than the TRANS-1 state. The TRANS-8 state occurs when the electric field is reapplied after the TRANS-6 state. Proteins are expected to be refocused from a normal concentration distribution. The steady state of carrier ampholytes after refocusing (SS-3) is similar but not identical to the SS-1 state because of the possible diffusion of certain ampholytes into the electrolytes during the TRANS-5 state and also because of the differences in the electric field strength, which may be intentional for experimental purposes. Similarly, the steady state of proteins after refocusing (SS-4) is similar but not identical to the SS-2 state for the same reasons described for the SS-3 state.

The kinetics of the focusing, steady state, deformation, defocusing, and refocusing of proteins can be evaluated directly by optical scanning of the species of interest.

C. Applications

A flow diagram of measurable parameters by TRANS-IF is shown in Fig. 1. The experimental data needed include changes in peak position ($\bar{x}$), peak variance σ^2, and peak area A as a function of time t. In conjunction with controllable variables such as polyacrylamide gel concentration and the use of pI markers, it is possible to obtain a number of relevant methodological parameters such as resolution R_s, resolving power ΔpI, minimal focusing time t_{MF}, segmental pH gradient $d(\text{pH})/dx$, and apparent physical constants of proteins, such as isoelectric point pI, diffusion coefficient D, retardation coefficient C_R, and the slope of the pH–mobility curve at pI, $dM/d(\text{pH})$. Practically, measurement of the minimal focusing time, the resolution, and the isoelectric point of proteins is of primary importance in an isoelectric focusing experiment. Minimally, we want to know whether a protein has been focused, what its isoelectric point is, and how well it is separated from other adjacent proteins in the pH gradient. TRANS-IF provides this type of information accurately and *in situ* in the form of a computer output.

II. INSTRUMENTAL ASPECTS

TRANS-IF experiments have been performed with three types of scanning instruments described in detail elsewhere (Catsimpoolas 1971a–c, 1975a; Catsimpoolas *et al.*, 1975). In general, the light source is a deuterium or xenon lamp with collimating ultraviolet optics. One or two quartz monochromators in tandem arrangement are used to select the wavelength. Rectangular slits of variable width (10–100 μm) placed behind the monochromator, in front of the photomultiplier, or both direct the light beam to a thin plane perpendicular to the length of the column. The column moves vertically (or horizontally with gels) at constant speed using either synchronous or stepping motors. The speed of scanning can be varied. The column bears electrolyte reservoirs which move along with the column, so that the electric field is applied at all times during scanning. The column is cylindrical; it is made of excellent quality quartz (Suprasil) for optical trans-

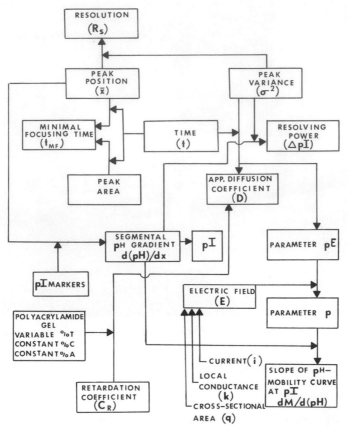

Fig. 1. Flow diagram of the relationship among various parameters and apparent physical constants measured in TRANS-IF. [From Catsimpoolas (1975a); used with permission of Plenum Press, New York.]

parency in the visible and ultraviolet range. In the two latter instruments (Catsimpoolas 1971b, Catsimpoolas *et al.*, 1975), the column is cooled to the desired temperature by circulating water from a thermostated water bath. The detection system consists of a photomultiplier, photometer, and a strip chart or $X-Y$ recorder. The photometer output is digitized by an analog-to-digital converter and punched on a paper tape. To date, data have been analyzed primarily off-line by a time-sharing remote computer. A more advanced multiple-column instrument coupled to a dedicated PDP/11 computer with graphic display capability and disk storage is under construction. A photograph of the third prototype (Catsimpoolas *et al.*, 1975) instrument described in the literature is shown in Fig. 2.

Fig. 2. Photograph of the TRANS-IF instrument. [From Catsimpoolas *et al.* (1975).]

III. DATA ACQUISITION AND PROCESSING

A. Data Acquisition

The analog signal from the photomultiplier during the scanning of the column is a continuous current (or voltage) whose amplitude is proportional to the absorbance being measured. In order to be processed by a computer the analog voltage must be digitized to produce a word whose number of bits after conversion is equivalent to the input voltage. To avoid large errors, the maximum input voltage has to be amplified to the maximum voltage that the analog-to-digital converter can accommodate, e.g., 10 V. A digital voltmeter can measure voltages as low as a few microvolts with as little as 0.01% error, but analog-to-digital conversion requires at least 10 msec for completion. In practice, this means that we can obtain a maximum of 100 data points from a peak 0.1 cm wide (at the base) at a scanning speed of 10 sec/cm. Such slow converters act as a low pass filter and thereby smooth the data to a considerable extent. However, fast analog-to-digital converters can process an 8-bit word in less than 10 μsec, which is indeed much faster than is normally required for scanning isoelectric focusing columns. In this case, amplification of the input signal is usually necessary.

In TRANS-IF experiments the analog signal produced by the photometer is digitized by an analog-to-digital converter at preselected time intervals. Since the mechanical scanning speed v of the column is constant, the digitized values represent the photometer signal amplitude at equidistant points (i.e., center of the window width formed by the slit). This is practically possible only if the rate of analog-to-digital conversion is very fast in comparison to the mechanical transport of the column. The number of data points j obtained per peak depends on the velocity v (centimeters per second) of the scanner, the digitizing rate g (data points per second), and the total width w_t (centimeters) of the peak profile at the baseline, so that

$$j = gw_t/v \tag{1}$$

To a first approximation, the narrower the slit width(s), the more accurately the peak profile will be recorded. However, the narrower the slit width, the larger is the random noise in the signal. A large slit width smooths the data by averaging the light absorbance over a larger window. In practice, the slit width should be approximately equal to the ratio w_t/j. Depending on baseline noise and peak shape (Catsimpoolas and Griffith, 1973) the desired value of j can be set between 10 and 100. If the peaks are of Gaussian shape, 10 data points may be sufficient to obtain their statistical moments. The more asymmetric the peak, the more data points are needed for its characterization. The total number of points k to be acquired per scan depends also on the peak capacity n which is defined as the maximum number of peaks resolvable in the column, so that

$$k = jn \tag{2}$$

The peak capacity depends primarily on the average standard deviation $\bar{\sigma}$ (centimeters) of typical peaks. Since adjacent peaks can be considered completely resolved if the peak heights are at a distance of $4\bar{\sigma}$ from each other, the value of n can be approximated by $L/4\bar{\sigma}$, where L (centimeters) is the length of the column.

After acquisition, the digitized data can be stored on punched paper tape, magnetic tape, disks, or other data storage devices. The speed at which these devices store data must be compatible with the data acquisition rate.

B.   Smoothing of Data

The primary information from a scanning frame should be in the form of absorbance or percent transmittance of the compounds of interest versus the distance in the column. However, random errors which are characteristically described as noise are often superimposed on this information in an indistinguishable manner. Removal of the noise without degrading the underlying information can be performed by convoluted least-squares differentiation procedures (Grushka, 1975). Two important restrictions have to be taken into consideration in this type of analysis. First, the digitized data points must be at a fixed, uniform interval in the chosen abscissa (i.e., distance or time). In other words, each data point must be obtained at the same time interval from each preceding point. Second, the curves formed by graphing the points must be continuous and more or less smooth. All of the errors are assumed to be in the ordinate (absorbance, percent transmittance) and none in the abscissa.

The least-squares calculations are carried out by convolution of the data points with properly chosen sets of integers. In this procedure, we take a fixed number of points and evaluate the central point by a convoluting function (i.e., quadratic) followed by normalization. Next, the point at one end of the group is dropped, the next point at the other end is added, and the process is repeated. Table III contains the convolution integers and their normalizing factors for smoothing (zeroth derivative) polynomials of degrees 2 and 3 from 5 to 15 fixed numbers of points. As an example, the equation for

TABLE III

Smoothing (Zero-Derivative) Convolution Integers and
Normalizing Factors (Quadratic or Cubic)

Data point	Fixed number of points					
	15	13	11	9	7	5
−07	−78					
−06	−13	−11				
−05	42	0	−36			
−04	87	9	9	−21		
−03	122	16	44	14	−2	
−02	147	21	69	39	3	−3
−01	162	24	84	54	6	12
00	167	25	89	59	7	17
01	162	24	84	54	6	12
02	147	21	69	39	3	−3
03	122	16	44	14	−2	
04	87	9	9	−21		
05	42	0	−36			
06	−13	−11				
07	−78					
Normalizing factor	1105	143	429	231	21	35

smoothing a five-point quadratic or cubic convolute is

$$y = (-3y_{-02} + 12y_{-01} + 17y_{00} + 12y_{01} - 3y_{02})/35 \qquad (3)$$

The number of fixed points that may be used for the smoothing depends on how accurately the polynomial describes the experimental curve under examination. The number of points should be chosen so that no more than one inflection in the observed data is included in any convolution interval. Data digitized at high intensities (i.e., taken very close together) offer more flexibility in the choice of the number of points to be used. Generally, the more data points included in the convoluting function, the less is the noise, since this is reduced approximately as the square root of the number of points involved. Thus, a nine-point smoothing produces approximately a threefold improvement in the signal-to-noise ratio. Although there is no way to assess the degree of distortion of recorded data introduced by the smoothing procedure, experience gained with separation of known compounds may suggest compromise conditions between rate of digitizing and number of data points involved in the convolution.

C. Slope Analysis

In isoelectric focusing where the peak shape can be approximated by a Gaussian curve, slope analysis can be very useful in detecting the baseline and peak shoulders, and in estimating the resolution of strongly overlapping Gaussian peaks (Grushka, 1975). Slope analysis can be performed by convoluted least-squares differentiation procedures (Savitzky and Golay, 1964) similar to those used for smoothing. For most work it is adequate to estimate the first- and second-derivative convolutes. Tables IV and V list the convolution integers and their normalizing factors for first- and second-derivative calculations for polynomials of degrees 2 and 3 from 5 to 15 fixed numbers of points. As an example, the equation for deriving the first derivative by a five-point quadratic convolute is

$$y' = (-2y_{-02} - y_{-01} + 0y_{00} + y_{01} + 2y_{02})/10 \qquad (4)$$

and for deriving the second derivative by a seven-point convolute is

$$y'' = (5y_{-3} + 0y_{-02} - 3y_{-01} - 4y_{00} - 3y_{01} + 0y_{02} + 5y_{03})/42 \quad (5)$$

The first derivative of a Gaussian curve has a maximum ($+$) and a minimum ($-$) at the inflection points of the curve and passes through zero (0) at peak maximum. Even in the presence of extreme baseline noise, the inflection points can be located. The distance between the maximum or minimum and the zero position corresponds to the standard deviation (σ_1, σ_2) on the front and back side of the curve, respectively. For a Gaussian

TABLE IV

First-Derivative Convolution Integers and Normalizing Factors (Quadratic)

Data point	Fixed number of points					
	15	13	11	9	7	5
−07	−7					
−06	−6	−6				
−05	−5	−5	−5			
−04	−4	−4	−4	−4		
−03	−3	−3	−3	−3	−3	
−02	−2	−2	−2	−2	−2	−2
−01	−1	−1	−1	−1	−1	−1
00	0	0	0	0	0	0
01	1	1	1	1	1	1
02	2	2	2	2	2	2
03	3	3	3	3	3	
04	4	4	4	4		
05	5	5	5			
06	6	6				
07	7	7				
Normalizing factor	280	182	110	60	28	10

TABLE V

Second-Derivative Convolution Integers and Normalizing Factors (Quadratic or Cubic)

Data point	Fixed number of points					
	15	13	11	9	7	5
−07	91					
−06	52	22				
−05	19	11	15			
−04	−8	2	6	28		
−03	−29	−5	−1	7	5	
−02	−44	−10	−6	−8	0	2
−01	−53	−13	−9	−17	−3	−1
00	−56	−14	−10	−20	−4	−2
01	−53	−13	−9	−17	−3	−1
02	−44	−10	−6	−8	0	2
03	−29	−5	−1	7	5	
04	−8	2	6	28		
05	19	11	15			
06	52	22				
07	91					
Normalizing factor	6188	1001	429	462	42	7

distribution, $\sigma_1 = \sigma_2$. Thus, the first derivative can be utilized to detect the presence of an asymmetric peak ($\sigma_1 \neq \sigma_2$). In addition, the values of σ_1 and σ_2 can produce the higher statistical moments, skewness, and excess of an asymmetric bi-Gaussian distribution according to the equations:

$$m_0 = \frac{H(2\pi)^{1/2}}{2}(\sigma_1 + \sigma_2) \tag{6}$$

$$m_1' = x_m + \frac{2(\sigma_2 - \sigma_1)}{(2\pi)^{1/2}} \tag{7}$$

$$m_1 = \frac{2(\sigma_2 - \sigma_1)}{(2\pi)^{1/2}} \tag{8}$$

$$m_2 = -\frac{2(\sigma_2 - \sigma_1)^2}{\pi} + \frac{\sigma_2{}^3 + \sigma_1{}^3}{\sigma_2 + \sigma_1} \tag{9}$$

$$m_3 = \frac{8(\sigma_2 - \sigma_1)^3}{\pi(2\pi)^{1/2}} - \frac{6(\sigma_2 - \sigma_1)(\sigma_2{}^3 + \sigma_1{}^3)}{(2\pi)^{1/2}(\sigma_2 + \sigma_1)} + \frac{4(\sigma_2{}^4 - \sigma_1{}^4)}{(2\pi)^{1/2}(\sigma_2 + \sigma_1)} \tag{10}$$

$$m_4 = -\frac{12(\sigma_2 - \sigma_1)^4}{\pi^2} + \frac{24(\sigma_2 - \sigma_1)^2(\sigma_2{}^3 + \sigma_1{}^3)}{2\pi(\sigma_2 + \sigma_1)}$$

$$+ \frac{-32(\sigma_2 - \sigma_1)(\sigma_2{}^4 - \sigma_1{}^4) + 6\pi(\sigma_2{}^5 - \sigma_1{}^5)}{2\pi(\sigma_2 + \sigma_1)} \tag{11}$$

$$S = m_3/m_2{}^{3/2} \tag{12}$$

$$E = m_4/m_2{}^2 - 3 \tag{13}$$

where H is the peak height and x_m is the abscissa position of peak maximum from an arbitrary origin. In the case of a Gaussian curve, the σ estimated by slope analysis is useful in obtaining the area A of the peak from

$$A = H\sigma(2\pi)^{1/2} \tag{14}$$

and also to derive the half-Gaussian distribution ordinates for peak simulation from

$$A_i = \text{antilog}\left[\log\frac{1}{(2\pi)^{1/2}} - \frac{1}{2}\left(\frac{x_i - m_i}{\sigma}\right)^2 \log e\right]\left(\frac{H}{0.39894}\right) \tag{15}$$

where $(x_i - m_i) = (\Sigma\Delta x)_i$, the sum of coordinate increments at the ith interval. When $(x_i - m_i)/\sigma = 0$, $A_i = H = 0.39894$.

The second-derivative curve of a Gaussian distribution produces a minimum $(-)$ at the peak maximum of the original curve and two maxima $(+)$. The magnitudes (ordinate values) of these extrema depend on the height of the peak and its variance (σ^2). However, the ratio of either maximum to the minimum (ordinates) is independent of the height and variance of the peak and in fact its value is $-2 \exp[-3/2] = -0.446$. The ratio between the two maxima is unity. Defining R_1 as the ratio of the maximum on the front side of the curve to the minimum, R_2 as the ratio of the maximum on the back side of the peak to the minimum, and R_3 as the ratio of the two maxima on the front and back sides, the following applications have been suggested by Grushka (1975). In a Gaussian system, any deviation from the extrema ratios of $-2 \exp[-3/2]$ and 1.0 immediately indicates the existence of double peaks. Second-derivative analysis is preferred to the first derivative because it is sensitive to slope changes when one of the two peaks in the composite is one-tenth or less than the height of the second. In the case of two overlapping equal peaks the first-derivative ratio (maximum to minimum) is always unity. While the ratio of two maxima (R_3) of the second derivative is unity when the overlapping peaks are identical, the two ratios R_1 and R_2 increase with the resolution between the two peaks. By employing calibration curves (Grushka, 1975) with suitable standards, the R_1, R_2, R_3 ratios can indicate not only the resolution between strongly overlapping peaks in the composite but also the ratio of their heights. With single peaks, the standard deviation (σ) can be estimated from

$$\pm\sigma = (x - x_0)/\sqrt{3} \tag{16}$$

where the x's are the abscissa positions of the extrema and x_0 is the coordinate of the Gaussian's maximum (center of gravity).

D. Peak Detection and Baseline Correction

First- and second-derivative slope analysis can be utilized for distinguishing a peak maximum ($y' = 0$, $y'' < 0$) from a trough ($y' = 0$, $y'' > 0$), baseline conditions ($y' = 0$, $y'' = 0$), or a peak ascent ($y' > 0$, $y'' > 0$ then $y'' < 0$) from a peak descent ($y' < 0$, $y'' < 0$ then $y'' > 0$). In practice, "zero" for the y' is represented by the range of two threshold values C_1 and C_2, and that for the y'' by two additional threshold parameters C_3 and C_4. Baseline conditions are met when $C_1 > y' > C_2$ and $C_3 > y'' > C_4$. The selection of the parameters C_1, C_2, C_3, and C_4 depends on the magnitude of baseline noise in relation to the peak height. At first, a desired limit of integration (l) is selected such that it represents any percentage (e.g., 0.01, 0.1, 1.0%) of the peak height H. This is called an H_l increment. C_1 and C_2 have the same value but exhibit positive and negative signs, respectively. The values of C_1 and C_2 for a defined H_l are obtained from the first derivative y' of consecutive data points (5–15, see Table IV), increasing (C_1) or decreasing (C_2) by one H_l increment. The larger the H_l increment, the higher are the values of C_1 and C_2. For example, if H_l is set at 0.1% of the peak height, baseline conditions are met within the threshold window $+C_1$ to $-C_2$ (computed for H_l of 0.1 H), even if the baseline ascends or descends by one H_l increment. Thus, the H_l values determine the sensitivity of detection of slope changes, which signify the start and end of peaks. It should be noted that detection of "peak start" by the slope increase method should be corrected by storing the data preceding the point of peak detection to set back the baseline. For a Gaussian curve, the baseline should be set back by approximately 0.3σ. Similar corrections apply to the back side of the peak where additional points are added on after the point of baseline detection. As an example, the equations for deriving the C_1 and C_2 parameters for a seven-point moving average are

$$C_{1,2} = \pm(-3(H_l) - 2(2H_l) - (3H_l) + 0(4H_l)$$
$$+ (5H_l) + 2(6H_l) + 3(7H_l))/28 \qquad (17)$$

It can be seen that if the limit of integration is set at 0.1% of H the parameters C_1 and C_2 have values of $\pm 1.0 \times 10^{-2}$. The second-derivative threshold parameters C_3 and C_4 are set empirically at such levels as to avoid false triggering of peak start and end by baseline noise. Some guidelines can be obtained by entering alternate or fluctuating values of $+H_l$, $-H_l$ into the second-derivative (y'') equations of variable consecutive data points (see Table V).

Once the baseline on both sides of the peak has been detected, the curve has to be corrected for an ascending or descending baseline by linear interpolation. The method is based on the assumption that the peak is superimposed on a baseline exhibiting a linear sloping continuum. Correction of the distribution is carried out according to the equation

$$y_{corr} = y_{obs} - [(y_e - y_s)/(x_e - x_s)]x_{obs} - [(y_s x_e - y_e x_s)/(x_e - x_s)] \qquad (18)$$

where y_{corr} is the corrected ordinate, y_{obs} is the observed ordinate, x_{obs} is the corresponding coordinate, y_s and y_e are the ordinates at peak start and end, and x_s and x_e are the corresponding coordinates.

E. Moment Analysis

After baseline correction, the nth statistical moment (m_n') of the concentration distribution $c(x)$ of a peak is estimated by

$$m_n' = \frac{\int x^n C(x)\, dx}{\int C(x)\, dx} \qquad (19)$$

Subsequently, the second, third, and fourth central moments, measured relative to m'_1, are obtained by

$$m_n = \frac{\int (x - m'_1)^n C(x)\, dx}{\int C(x)\, dx} \tag{20}$$

where x is the distance and C is the concentration.

The first moment m'_1 is the center of gravity of the concentration profile. It coincides with the peak maximum only if the peak is symmetrical. The second central moment m_2 is the peak variance σ^2. The square root of m_2 corresponds to the standard deviation σ of the concentration distribution which provides a measure of peak width. The third central moment m_3 is indicative of the direction and magnitude of peak asymmetry, whereas the fourth central moment m_4 is a measure of peak flatness as compared to a Gaussian shape. Negative values of m_3 indicate a fronting shape, and positive values, tailing peaks. The zeroth moment is the normalized area of the peak.

In addition, the second, third, and fourth central moments can be used to estimate the coefficients of skewness S and excess E by

$$S = m_3/m_2^{3/2} \tag{21}$$

$$E = (m_4/m_2^2) - 3 \tag{22}$$

The skew measures the asymmetry of the peak while the excess indicates deviation from a Gaussian shape in regard to flatness. The coefficients S and E are dimensionless quantities and therefore independent of the size of the peak (Grushka, 1975).

In practice, the moments are estimated by digitizing the absorbance A_i at fixed time intervals which correspond to a fixed distance Δx in the column during a scan. Starting from an arbitrary origin, the coordinate at the ith interval is x_i. The moments are estimated by

$$m_0 = \sum A_i \Delta x \tag{23}$$

$$m'_1 = \sum A_i x_i \Big/ \sum A_i \tag{24}$$

$$m_2 = \sum A_i (x_i - m'_1)^2 \Big/ \sum A_i \tag{25}$$

$$m_3 = \sum A_i (x_i - m'_1)^3 \Big/ \sum A_i \tag{26}$$

$$m_4 = \sum A_i (x_i - m'_1)^4 \Big/ \sum A_i \tag{27}$$

It is apparent that the exact description of the peak area, position, width, and shape by moment analysis should have fundamental applications in isoelectric focusing. The connection between the two is made by adopting mathematical models describing the isoelectric focusing process in regard to the mass transport as a function of time and also as affected by other physical factors such as molecular sieving, solute concentration, and medium viscosity.

IV.   METHODOLOGICAL PARAMETERS

A.   Evaluation of the Steady State

Knowledge of the peak position, area, and variance as a function of time in TRANS-IF allows the observation of useful clues concerning the isoelectric focusing process. For example, at the steady state the statistical moments of the peak should remain constant with time. Any disturbance of the steady state will be reflected in instability of peak position, area, and variance. Deformation of the pH gradient may cause changes in the variance of a focused peak which depends partly on $d(\text{pH})/dx$, i.e., the segmental pH gradient. Changes in peak position may also occur if the pH drift is accompanied by anodic migration of the focused proteins (Fig. 3) (Catsimpoolas, 1973c). The variance of a peak at the steady state may also be affected by

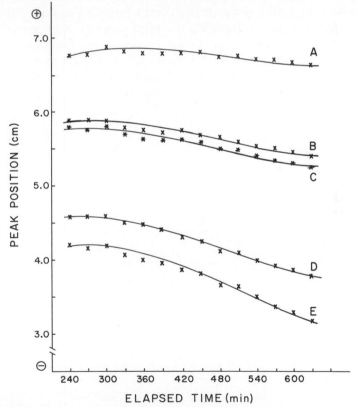

Fig. 3.   Peak position instability demonstrated in the TRANS-4 state in polyacrylamide gels. A, BB inhibitor; B, ovalbumin I; C, ovalbumin II; D, β-lactoglobulin A; E, β-lactoglobulin B. [From Catsimpoolas (1973c).]

association or dissociation of the protein at pI because of changes in the diffusion coefficient. Other instabilities may be due to temperature, viscosity, and conductance changes.

B. Minimal Focusing Time

Some of the factors that affect the minimal focusing time t_{MF} of proteins (i.e., the time required to migrate to their pI position) include the (a) pH–mobility curve of each protein, (b) sieving effects in gels, (c) presence of a viscosity gradient (e.g., sucrose), (d) carrier ampholyte concentration, (e) electric field strength, (f) pH range of carrier ampholytes, (g) temperature, (h) conductance course, (i) presence of additives (e.g., urea), and others. When a mixture of proteins is focused, each individual protein may reach its pI position at widely different times. It is therefore desirable to be able to determine the t_{MF} value of a protein either alone or in a mixture with other proteins. The latter case is obviously the most complex. An approach that has been utilized (Catsimpoolas, 1973c) is to measure the area of a discernible peak at the pI position until the steady state is achieved. The peak area should increase with time during focusing, attain a maximum value, and then remain constant at the steady state (Fig. 4). Such measurements

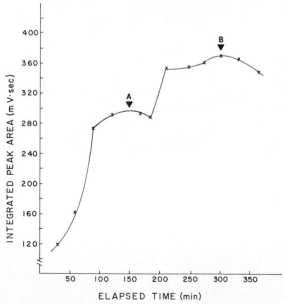

Fig. 4. Plot of integrated peak area versus time of focusing for ovalbumin in pH 3–10 (2%) Ampholine. Arrow B indicates the SS-2 state. [From Catsimpoolas (1973c).]

are facilitated if the protein is loaded in the uniform or gradient mode rather than in the pulse fashion. In the simple case of a single amphoteric species (e.g., histidyltyrosine) or protein, the position of the discernible peak(s) can be followed (see Figs. 5–7) until it remains constant at pI (Catsimpoolas et al., 1974a). If the compound has been loaded in the uniform mode, two discernible peaks migrate from the two ends of the column (positive and negative) until they merge into one at pI. In the pulse mode of loading, only one peak can be distinguished. The time at which constancy of pI position is attained is designated as t_{MF}. A typical example concerning the evaluation of the effect of carrier ampholyte concentration on the t_{MF} of histidyltyrosine in a standardized polyacrylamide gel system is shown in Fig. 8. Such experiments are needed to assess objectively the effect of other factors on the t_{MF} of ampholytes and proteins in typical isoelectric focusing systems.

C.   Resolution, Resolving Power, Segmental pH Gradient, and Isoelectric Point

The resolution (i.e., degree of separation) between two adjacent zones in isoelectric focusing can be expressed (Catsimpoolas, 1973e) as

$$R_s = \Delta \bar{x}/1.5(\sigma_A + \sigma_B) \tag{28}$$

where $\Delta \bar{x}$ (centimeters) is the peak separation between two zones A and B with standard deviations σ_A and σ_B (in centimeters). A resolution of unity signifies two just-resolved zones (Svensson, 1966).

The resolving power (Vesterberg and Svensson, 1966) is defined as

$$\Delta pI = 3[d(\mathrm{pH})/dx]\sigma \tag{29}$$

where ΔpI is the minimum pI difference for complete resolution of two focused zones and $d(\mathrm{pH})/dx$ is the segmental pH gradient. The latter parameter can be measured by using two pI markers of closely spaced isoelectric points (Catsimpoolas, 1973c) from

$$\Delta(\mathrm{pH})/\Delta x = (pI_A - pI_B)/(\bar{x}_A - \bar{x}_B) \tag{30}$$

where pI is the isoelectric point, $\bar{x}$ is the peak position, and subscripts A and B denote two pI markers. The assumption is made that species A and B have reached their isoelectric points and that $\Delta(\mathrm{pH})/\Delta x$ is constant between pI_A and pI_B, where $\bar{x}_A - \bar{x}_B$ represents a small segment of the separation path. Since $\bar{x}$ and σ can be obtained by TRANS-IF, the resolution, segmental pH gradient, and resolving power can be obtained directly as a function of time, electric field strength, and other factors (Catsimpoolas, 1973e).

The apparent isoelectric point of an "unknown" protein (U) can also be determined by TRANS-IF. If the protein is focused in the region of an

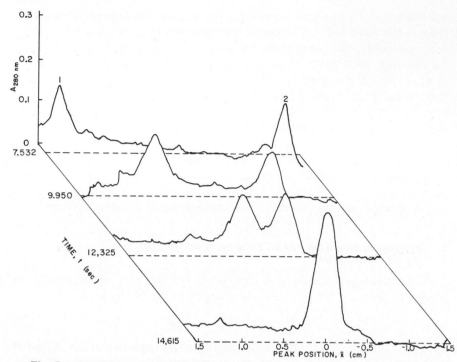

Fig. 5. Scanning electrophoretic patterns of histidyltyrosine peaks migrating toward the p*I* position. Peak 1 migrates away from the negative electrode and peak 2 from the positive. The p*I* position is arbitrarily set at $\bar{x} = 0$. The original sample distribution was uniform (UN); Ampholine concentration 2%. [From Catsimpoolas *et al.* (1974a); used with permission of Elsevier Publ. Co., Amsterdam.]

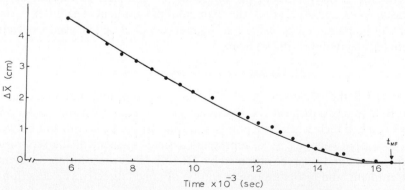

Fig. 6. Plot of peak position difference ($\Delta \bar{x}$) (see Fig. 5) versus time. Histidyltyrosine, 3% Ampholine concentration 2% [From Catsimpoolas *et al.* (1974a); used with permission of Elsevier Publ. Co., Amsterdam.]

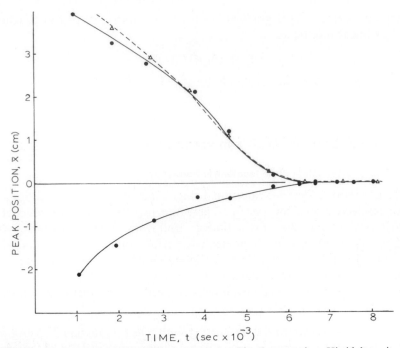

Fig. 7. Plot of peak position difference ($\Delta \bar{x}$) (see Fig. 5) versus time. Histidyltyrosine, 3% Ampholine concentration (UN sample loading). [From Catsimpoolas *et al.* (1974a); used with permission of Elsevier Publ. Co., Amsterdam.]

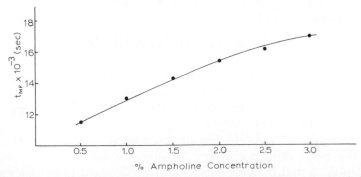

Fig. 8. Plot of peak position versus time. Soybean trypsin inhibitor, 2% Ampholine concentration. Solid circles represent UN sample loading, and open triangles PN loading. [From Catsimpoolas *et al.* (1974a); used with permission of Elsevier Publ. Co., Amsterdam.]

estimated "segmental pH gradient" as described above, its pI can be calculated by linear interpolation:

$$pI_U = pI_A + (\Delta(pH)/\Delta x)(\bar{x}_A - \bar{x}_U) \tag{31}$$

All three species A, B, and U should be at pH equilibrium, i.e., at the steady state.

V. KINETICS OF DEFOCUSING AND REFOCUSING

A. Theory of Transient State Isoelectric Focusing

As described earlier, two additional stages can be generated intentionally in an isoelectric focusing system. These are (a) defocusing, in which the electric field is abolished for a time t_2 (following focusing in which the system is allowed to approach the steady state for a time t_1) and (b) refocusing for a time t_3, in which the field is reapplied and the distribution again approaches the steady state. In the defocusing stage the focused zone is allowed to spread by diffusion. The advantages of performing kinetic experiments during the refocusing period are

(a)   focusing is carried out by starting with a near Gaussian distribution,
(b)   the zone is restricted to a narrow region of the pH (and therefore mobility) spectrum near the isoelectric point, and
(c)   data are collected under conditions of nearly linear pH gradient [$d(pH)/dx$] and linear pH–mobility relationship [$dM/d(pH)$].

Recently, Weiss et al. (1974) provided a restricted theory of the kinetics of focusing, defocusing, and refocusing. In order to permit a first theoretical approximation of a very complex system, certain assumptions were adopted which may be relaxed later to provide a more generalized theory. These assumptions are as follows:

1.   A linear and stable pH gradient is established prior to application of the sample protein.
2.   The pH–mobility curve of the protein is assumed to be linear. This assumption is valid only for a limited region near the isoelectric point. On the basis of these two assumptions, $p = dM/dx$ is constant.
3.   The electrical field strength E is assumed to be uniform throughout the entire separation path. In lieu of assumptions 1–3, we could simply assume that $pE = dv/dx$ is constant, where v is velocity.
4.   Diffusion and mobility coefficients are assumed to be independent of concentration.
5.   Diffusion coefficients are assumed to be independent of pH (at least in the region near the isoelectric point).

6. It is assumed that there are no physical/chemical interactions between the protein and other chemical species present (e.g., ampholytes), and no self-association or protein–protein interactions.

7. Band spreading is governed only by diffusion or by a diffusion-like process. Thus electrostatic effects are ignored and it is assumed that the protein is perfectly homogeneous with respect to pI, charge, mobility, radius, and diffusion coefficient.

8. No perturbing phenomena such as electroendosmosis, convective disturbances, or precipitation at the isoelectric point are present.

9. If a gel or density gradient is used as a supporting medium, their effects on diffusion coefficients and on mobility are negligible (or at least constant throughout the gel), and there is no effect on the uniformity of the electrical field. Thus the effect of the viscosity gradient which is superimposed on the density gradient in sucrose–gradient columns is ignored. Likewise, the molecular sieving effects that are present when polyacrylamide gels are used as a supportive medium are ignored.

10. The effect of the boundary condition that there can be no flux of the species of interest through the ends of the gel column, or that there is no abrupt discontinuity of pH at the ends of the column, is ignored. These effects should become insignificant shortly after the start of the experiment.

For experimental purposes it is sufficient and convenient to find $\mu_1(\tau)$ and $\sigma^2(\tau)$, i.e., the mean and square of the standard deviation of peak width. These equations are

1. Focusing ($0 \leq \tau \leq \tau_1$):

$$\mu_1(\tau) = \mu_1(0)e^{-\tau} + y_0(1 - e^{-\tau}) \qquad (32)$$

$$\sigma^2(\tau) = \sigma^2(0)e^{-2\tau} + \alpha(1 - e^{-2\tau}) \qquad (33)$$

2. Defocusing ($\tau_1 \leq \tau \leq \tau_1 + \tau_2$):

$$\mu_1(\tau) = \mu(0)e^{-\tau_1} + y_0(1 - e^{-\tau_1}) = \text{const} \qquad (34)$$

$$\sigma^2(\tau) = \sigma^2(0)e^{-2\tau_1} + \alpha(1 - e^{-2\tau_1}) + 2\alpha(\tau - \tau_1) \qquad (35)$$

3. Refocusing ($\tau_1 + \tau_2 \leq \tau$):

$$\mu_1(\tau) = \{\mu_1(0)e^{-\tau_1} + y_0(1 - e^{-\tau_1})\} \exp\{-(\tau - \tau_1 - \tau_2)\}$$
$$+ y_0(1 - \exp\{-(\tau - \tau_1 - \tau_2)\}) \qquad (36)$$

$$\sigma^2(\tau) = \{\sigma^2(0)e^{-2\tau_1} + \alpha(1 - e^{-2\tau_1}) + 2\alpha\tau_2\} \exp\{-2(\tau - \tau_1 - \tau_2)\}$$
$$+ \alpha(1 - \exp\{-2(\tau - \tau_1 - \tau_2)\}) \qquad (37)$$

where L is the column length, x_0 is the position of the isoelectric point, $\alpha = D/(L^2\, pE)$, $\tau = pEt$, and $y_0 = x_0/L$.

A computer simulation study derived from theory of the time course of the centroid μ and σ^2 in TRANS-IF is shown in Fig. 9.

The centroid approaches the isoelectric point by an exponential decay during focusing and refocusing. With ideal initial pulse loading, the bandwidth σ^2 increases during focusing, asymptotically approaching the steady state value.

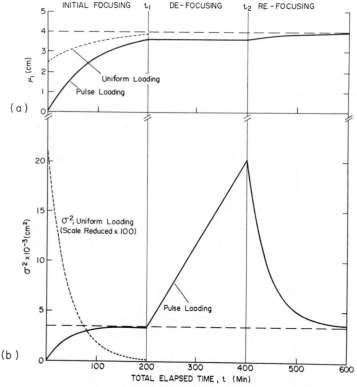

Fig. 9. Effect of Ampholine concentration on minimal focusing time (t_{MF}) of histidyl-tyrosine. [From Catsimpoolas et al. (1974a); used with permsiion of Elsevier Publ. Co., Amsterdam.]

The greatest departure from the assumptions is expected to occur during the focusing period. For this reason, only data obtained during the de-focusing and refocusing period can be utilized to validate the theory. The use of a narrow segment of the separation path minimizes errors due to unequal distribution of ampholytes which may lead to a nonuniform conductance and viscosity course. However, a major problem is the accurate estimation of the field strength E which depends on the conductance course throughout the pH gradient. The problem cannot be solved without the availability of suitable carrier ampholytes which would provide linear and stable pH gradients and uniform conductance, viscosity, and concentration throughout the column.

B. Measurement of D and pE

Experimentally, the kinetics of defocusing and refocusing can be determined by following the changes of σ^2, which is the square of the standard deviation of peak width, versus elapsed time. The equations describing the behavior of σ^2 during these two stages of the experiment have been derived from theory to be

1. Defocusing

$$\sigma^2(t_2) = \sigma^2(t_1) + 2D(t_2 - t_1) \tag{38}$$

2. Refocusing

$$\sigma^2(t) = (D/pE) + 2Dt_2 \exp(-2pEt_3) \tag{39}$$

Experimentally, a plot of σ^2 versus $2t_2$ should permit estimation of the apparent diffusion coefficient D (as the slope of the line) during the defocusing state. Also, a plot of $\log_{10}[(\sigma_R^2 - \sigma_F^2)/\sigma_D^2]$ versus $2t_3$ during the refocusing stage can be used to determine the parameter $-pE/2.303$ as the slope of the linear plot. If $d(pH)/dx$ and E are known, the physical constant $dM/d(pH)$ can be estimated from

$$dM/d(pH) = p/[d(pH)/dx] \tag{40}$$

where $\sigma_F^2 = D/pE$ is the variance at the steady state; σ_R^2 is the variance at any time of refocusing; σ_D^2 is the variance at the end of defocusing, and

$$E = i/q\kappa \tag{41}$$

where i is the current, q is the cross-sectional area, and κ is the conductance of the field.

Representative concentration profiles of an amphoteric compound, histidyltyrosine, during defocusing and refocusing are shown in Fig. 10. A typical time course of the variance σ^2 as a function of time starting with the onset of defocusing is shown in Fig. 11. During defocusing σ^2 increases in a linear fashion with time whereas during refocusing σ^2 is satisfactorily described by a negative exponential decay function, as expected from theory. A linearized plot of the exponential decay during refocusing is depicted in Fig. 12. Detailed aspects of the kinetics of defocusing and refocusing as a function of sample load and ampholyte concentration have been given by Catsimpoolas et al. (1974b). The apparent diffusion coefficient D depends on

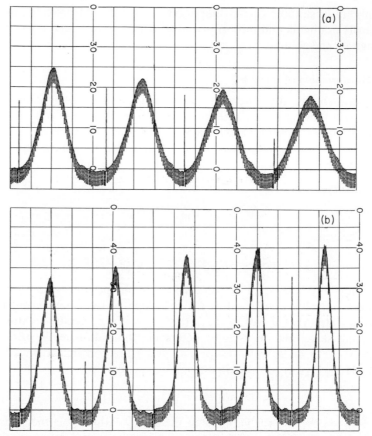

Fig. 10. TRANS-IF spectra of histidyltyrosine during the stages of defocusing (a) and re-focusing (b). The event marks under the continuous curve indicate the time at which absorbance readings were digitized. [From Catsimpoolas *et al.* (1974b).]

zone load in approximately a linear fashion and it is therefore possible to extrapolate to zero load by linear regression techniques. Similarly, the parameter pE is zone load dependent. The lowest values of D and pE are obtained at zero load, as expected. The effect of carrier ampholyte concentration on D and pE is more difficult to explain. Increasing ampholyte concentration causes a decrease in apparent D and an increase in apparent pE, i.e., increased mobility at the region of pI. Although increase in the viscosity of the system due to the presence of the ampholyte could explain the decrease in D, this is not compatible with the increase in mobility. Again, this may be the result of nonideal effects on E due to the uneven ampholyte course.

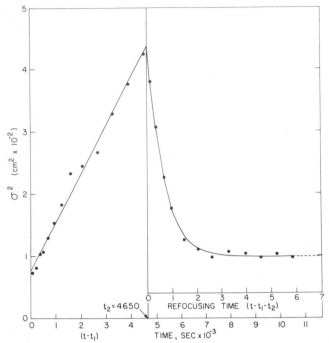

Fig. 11.   Plot of corrected second moment (σ^2) for histidyltyrosine versus elapsed time during defocusing, refocusing, and the refocusing steady state (experimental results). [From Catsimpoolas *et al.* (1974b).]

From the aforementioned, it can be concluded that although TRANS-IF can provide the necessary physicochemical measurements, the unsuitability of the present ampholytes makes it very difficult to obtain corrected values of D and $dM/d(\text{pH})$. This will have to await the preparation of "second generation" ampholytes suitable for physicochemical studies. If such ampholytes were available, we should also be able to obtain the complete pH–mobility curve of a protein from one experiment as discussed and partly demonstrated by Catsimpoolas *et al.* (1974a).

C.   Measurement of the Retardation Coefficient C_R in Polyacrylamide Gels

The apparent diffusion coefficient in polyacrylamide gels is related to gel concentration (Rodbard and Chrambach, 1970) by

$$D_T = D_0 \exp(-bT) \tag{42}$$

where D_0 is the free diffusion coefficient, T is the gel concentration, D_T is the diffusion coefficient at any gel concentration T, and b is the retardation

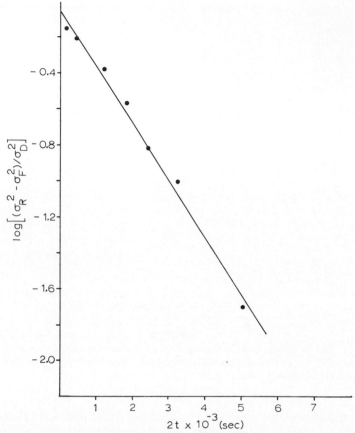

Fig. 12. Kinetics of refocusing of histidyltyrosine. The slope of the line corresponds to $-pE/2.303$. [From Catsimpoolas (1975b).]

coefficient $(C_R = -d \log_{10}(D/dT)$ multiplied by 2.303. The retardation coefficient C_R obtained from diffusion data (defocusing stage) should be analogous to the retardation coefficient K_R (Rodbard and Chrambach, 1970) derived from electrophoretic mobility data. Equation (42) can be written as

$$\log(D_T/D_0) = -C_R T \qquad (43)$$

A plot of $\log D_T$ or $\log(D_T/D_0)$ versus T will produce C_R as the slope of the line.

Measurements of C_R have been performed from the defocusing stage in variable %T polyacrylamide gels (Catsimpoolas, 1973d). A linear relationship was obtained between $\log D_T$ and %T. However, the extrapolated D_0 values were much higher than expected. This phenomenon has been observed before (Lunney *et al.*, 1971) but cannot be explained. Thus, the C_R and D_0 values obtained by TRANS-IF may be used only as apparent constants for relative comparison of molecular size at pI.

VI. REMARKS

Despite the considerable success of isoelectric focusing in the separation of proteins that exhibit minor charge differences, the method remains largely empirical in practice. The TRANS-IF technique is capable of providing the necessary measurements for the much needed transition into the quantitative realm, but it is hampered by the nonideal behavior of the currently available carrier ampholytes. Nevertheless, TRANS-IF is still useful in the determination of important methodological parameters such as minimal focusing time, resolution, and tests for the attainment or deformation of the steady state. It is hoped that the development of a commercial apparatus in the near future will make the technique available to other investigators and, thus, escape its possible labeling as a laboratory curiosity.

ACKNOWLEDGMENTS

This work was supported by NSF grant MPS74-19830 and NCI contract No. NO1-CB-43928.

REFERENCES

Catsimpoolas, N. (1971a). *Separ. Sci.* **6**, 435–442.
Catsimpoolas, N. (1971b). *Anal. Biochem.* **44**, 411–426.
Catsimpoolas, N. (1971c). *Anal. Biochem.* **44**, 427–435.
Catsimpoolas, N. (ed.) (1975a). "Methods of Protein Separation," Vol. 1, pp. 27–67. Plenum, New York.
Catsimpoolas, N. (1975b). *In* "Isoelectric Focusing of Proteins and Related Substances" (J. P. Arbuthnott, ed.), pp. 58–73. Butterworths, London.
Catsimpoolas, N. (1973a). *Ann. N.Y. Acad. Sci.* **209**, 65–79.
Catsimpoolas, N. (1973b). *Fed. Proc.* **32**, 625.
Catsimpoolas, N. (1973c). *Anal. Biochem.* **54**, 66–78.
Catsimpoolas, N. (1973d). *Anal. Biochem.* **54**, 79–87.
Catsimpoolas, N. (1973e). *Anal. Biochem.* **54**, 88–94.
Catsimpoolas, N., and Campbell, B. E. (1972). *Anal. Biochem.* **46**, 647–676.
Catsimpoolas, N., and Griffith, A. L. (1973). *Anal. Biochem.* **56**, 100–120.
Catsimpoolas, N., and Wang, J. (1971). *Anal. Biochem.* **39**, 141–155.
Catsimpoolas, N., Campbell, B. E., and Griffith, A. L. (1974a). *Biochim. Biophys. Acta* **351**, 196–204.
Catsimpoolas, N., Yotis, W. W., Griffith, A. L., and Rodbard, D. (1974b). *Arch. Biochem. Biophys.* **163**, 113–121.

Catsimpoolas, N., Griffith, A. L., Williams, J. M., Chrambach, A., and Rodbard, D. (1975). *Anal. Biochem.* **69**, 372–384.

Grushka, E. (1975). *In* "Methods of Protein Separation" (N. Catsimpoolas, ed.), Vol. 1, pp. 161–192. Plenum, New York.

Lundahl, P., and Hjertén, S. (1973). *Ann. N.Y. Acad. Sci.* **209**, 94–111.

Lunney, J., Chrambach, A., and Rodbard, D. (1971). *Anal. Biochem.* **40**, 158–173.

Rilbe, H. (1973). *Ann. N.Y. Acad. Sci.* **209**, 80.

Rodbard, D., and Chrambach, A. (1970). *Proc. Nat. Acad. Sci. U.S.* **85**, 970–977.

Savitzky, A., and Golay, M. J. E. (1964). *Anal. Chem.* **36**, 1627–1639.

Svensson, H. (1961). *Acta Chem. Scand.* **15**, 325–341.

Svensson, H. (1966). *J. Chromatogr.* **25**, 266–273.

Vesterberg, O., and Svensson, H. (1966). *Acta Chem. Scand.* **20**, 820–834.

Weiss, G. H., Catsimpoolas, N., and Rodbard, D. (1974). *Arch. Biochem. Biophys.* **163**, 106–112.

INDEX